# Military Construction and Appropriations

## Policy, Oversight, and Budgeting in the U.S. Defense Sector

snova
New York

# Veterans: Benefits, Issues, Policies and Programs Series

**Supporting Our Veterans: Comprehensive Care for Mental Health and Substance Use Disorders**
Gina P. Sharpe (Editor)
ISBN: 979-8-89530-504-1
eBook ISBN: 979-8-89530-522-5

**Health Care for Veterans: Background, Education, System Reform and Improvements**
Juliette Binet (Editor)
ISBN: 978-1-53617-530-1
eBook ISBN: 978-1-53617-531-8

**The Veterans Choice Program (VCP): Access, Implementation and Improvements**
Mabel Page (Editor)
ISBN: 978-1-53614-819-0
eBook ISBN: 978-1-53614-820-6

**Veterans Affairs Information Technology: Progress, Modernization and Challenges**
Bryan Barton (Editor)
ISBN: 978-1-53616-452-7
eBook ISBN: 978-1-53616-453-4

**Veterans: Benefits, Issues, Policies and Programs. Volume 6**
Timothy C. Roberts (Editor)
ISBN: 978-1-53613-825-2
eBook ISBN: 978-1-53613-826-9

**Veterans: Benefits, Issues, Policies and Programs. Volume 5**
Timothy C. Roberts (Editor)
ISBN: 978-1-63483-751-4
eBook ISBN: 978-1-63483-752-1

**Veterans: Benefits, Issues, Policies and Programs. Volume 4**
Timothy C. Roberts (Editor)
ISBN: 978-1-63321-722-5
eBook ISBN: 978-1-63321-752-2

**More information about this series can be found at**
https://novapublishers.com/product-category/series/veterans-benefits-issues-policies-and-programs/

# Military Construction and Appropriations

## Policy, Oversight, and Budgeting in the U.S. Defense Sector

snova
New York

# Veterans: Benefits, Issues, Policies and Programs Series

**Supporting Our Veterans: Comprehensive Care for Mental Health and Substance Use Disorders**
Gina P. Sharpe (Editor)
ISBN: 979-8-89530-504-1
eBook ISBN: 979-8-89530-522-5

**Health Care for Veterans: Background, Education, System Reform and Improvements**
Juliette Binet (Editor)
ISBN: 978-1-53617-530-1
eBook ISBN: 978-1-53617-531-8

**The Veterans Choice Program (VCP): Access, Implementation and Improvements**
Mabel Page (Editor)
ISBN: 978-1-53614-819-0
eBook ISBN: 978-1-53614-820-6

**Veterans Affairs Information Technology: Progress, Modernization and Challenges**
Bryan Barton (Editor)
ISBN: 978-1-53616-452-7
eBook ISBN: 978-1-53616-453-4

**Veterans: Benefits, Issues, Policies and Programs. Volume 6**
Timothy C. Roberts (Editor)
ISBN: 978-1-53613-825-2
eBook ISBN: 978-1-53613-826-9

**Veterans: Benefits, Issues, Policies and Programs. Volume 5**
Timothy C. Roberts (Editor)
ISBN: 978-1-63483-751-4
eBook ISBN: 978-1-63483-752-1

**Veterans: Benefits, Issues, Policies and Programs. Volume 4**
Timothy C. Roberts (Editor)
ISBN: 978-1-63321-722-5
eBook ISBN: 978-1-63321-752-2

**More information about this series can be found at**
https://novapublishers.com/product-category/series/veterans-benefits-issues-policies-and-programs/

**George K. Santiago**
Editor

# Military Construction and Appropriations

## Policy, Oversight, and Budgeting in the U.S. Defense Sector

snova
New York

Additional color graphics may be available in the e-book version of this book.

**Library of Congress Cataloging-in-Publication Data**

ISBN: 979-8-89530-799-1 (Hardcover)
ISBN: 979-8-89530-906-3 (eBook)

*Published by Nova Science Publishers, Inc. † New York*

# Contents

# Preface

The Military Construction (MILCON) program enables the Department of Defense to plan, program, design, and construct infrastructure to support U.S. military forces both domestically and abroad. The MILCON process involves a close partnership between DOD and Congress to build military facilities, starting with the identification of new infrastructure needs and culminating in the completion of the facility. This book explains the MILCON process, delves into the various federal laws that govern it, and highlights some of the legal and institutional challenges that can arise along the way.

# Chapter 1

# Military Construction: Authorities and Processes*


## Andrew Tilghman



## Summary

Congress appropriates several billion dollars annually to support construction projects for the Department of Defense's (DOD's) worldwide military installations. The military construction (MILCON) program enables the DOD to plan, program, design, and build infrastructure including runways, piers, warehouses, barracks, schools, hospitals, child development centers, and other facilities intended to support U.S. military forces at home and abroad. The MILCON process involves DOD and Congress acting together to build military facilities, beginning with development of new military infrastructure requirements and culminating in a completed facility. The lengthy and complex process can take five to seven years or more. Summarized, the MILCON process encompasses multiple steps, to include

- Identifying the need for a new facility, which can involve input from installation commanders, military department headquarters, weapon system program offices, and combatant commanders;
- Prioritizing construction projects within the military chain of command for the military departments or other DOD components;


---

* This is an edited, reformatted and augmented version of Congressional Research Service Publication No. R44710, updated September 5, 2024.


In: Military Construction and Appropriations
Editor: George K. Santiago
ISBN: 979-8-89530-799-1

- Consolidation and budgeting within the Office of the Secretary of Defense to create the military construction portion of the annual budget request and Future Years Defense Program (FYDP);
- Finalizing the annual President's budget request to Congress with a final list of projects seeking funding for the current budget year;
- Reviewing and adjusting the list of projects by congressional defense committees;
- Passage and enactment of authorization and appropriation acts; and
- Design and execution of the construction projects by a designated DOD construction agent.

## Introduction

The military missions of the Department of Defense (DOD) and the three military departments— the Army, the Navy, and the Air Force—drive requirements for facilities and infrastructure at military installations around the world.[1] As missions and technologies change, or as organizations move to or away from installations, the departments may need to build or dispose of facilities and infrastructure to meet new requirements. Similarly, the departments may need to replace, reconfigure, or modernize facilities and supporting infrastructure as buildings age or become obsolete.

While each of the military departments has a unique process for vetting and setting MILCON priorities, in general the process conforms with DOD's Planning, Programming, Budgeting, and Execution (PPBE) process for resource allocation.[2] Most projects can take three to five years or more to move from the time when senior military officials identify the need for construction of a new facility (i.e., a requirements determination) through the planning,

---

[1] Military departments include the Department of the Army, Department of the Navy (including the Navy and Marine Corps), and the Department of the Air Force (including the Air Force and the Space Force).

[2] Exceptions include DOD authorities under 10 U.S.C. §§2803, 2804, 2808, and 2854, which are generally not part of the PPBE process. See DOD, Department of Defense Directive (DODD) 4270.5, *Military Construction*, updated August 31, 2018, p. 3, at https://www.esd.whs.mil/Portals/54/Documents/DD/issuances/dodd/427005p.pdf. For more information on the PPBE process, see CRS In Focus IF10429, *Defense Primer: Planning, Programming, Budgeting, and Execution (PPBE) Process*, by Brendan W. McGarry and CRS Report R47178, *DOD Planning, Programming, Budgeting, and Execution (PPBE): Overview and Selected Issues for Congress*, by Brendan W. McGarry.

programming, and budgeting processes. When adding the time required for congressional approval, implementation of the federal contracting process (e.g., soliciting bids, awarding a contract), and executing the construction project, the overall process could span five to seven years or more.

This report describes and explains the life cycle of the MILCON process involving both DOD and Congress. The cycle typically begins with DOD identifying a new construction requirement and developing cost estimates based on preliminary planning and design work. It involves Congress' reviewing, adjusting and approving military construction plans. After the President signs into law the authorizations and appropriations for new military construction projects, the DOD and military departments manage and execute those projects. This report provides details about the various federal statutes that govern the process, and outlines some of the legal and institutional challenges that can arise. Typically, military construction (MILCON) accounts for about than 2% of the total DOD budget.[3]

## The Legal and Budgetary Framework

Title 10, Chapter 169, of the *United States Code* (hereinafter *U.S. Code* or U.S.C.) is titled, "Military Construction and Military Family Housing," and contains most of the provisions governing military construction. Section 2801 of Chapter 169 provides definitions of certain terms used throughout the chapter and elsewhere in law. In general, military construction applies to all buildings, structures, training ranges, and other improvements to real property that are located on a military installation.[4] The term *military installation* is

---

[3] 3 The DOD portion of the FY2023 President's budget request included a total of $12.2 billion for MILCON accounts (including family housing), about 1.6% of the department's total requested discretionary budget authority of $773 billion. For more information, see DOD, Office of the Under Secretary of Defense (Comptroller)/Chief Financial Officer, *Defense Budget Overview: United States Department of Defense Fiscal Year 2023 Budget Request*, Table A-1, "DoD Total (Base + Supplemental) Budget by Appropriation Title," April 2022, at https://comptroller.defense.gov/Portals/45/Documents/defbudget/FY2023/FY2023_Budget_Request_Overview_Book. pdf#page=135.

[4] See10 U.S.C. §2801.The statute defines the term military construction as "any construction, development, conversion, or extension of any kind carried out with respect to a military installation, whether to satisfy temporary or permanent requirements, or any acquisition of land or construction of a defense access road." The statute also specifies that a MILCON project include "all military construction work ... necessary to produce a complete and usable facility or a complete and usable improvement to an existing facility (or to produce

defined in 10 U.S.C.§2801(c)(4) to mean "a base, camp, post, station, yard, center, or other activity under the jurisdiction of the Secretary of a military department or, in the case of an activity in a foreign country, under the operational control of the Secretary of a military department or the Secretary of Defense, without regard to the duration of operational control."[5]

For the reserve components, Chapter 1803 of Title 10, U.S.C., "Facilities for Reserve Components," provides additional information pertaining to joint construction, real property exchange, and other items related to Reserve Component facilities.

In prescribing the various MILCON authorities and responsibilities, Chapter 169 regularly uses the term *Secretary concerned*, referring to the Secretaries of the military departments, or the Secretary of Defense in the case of a defense agency.[6]

Each year, the DOD submits a detailed list of MILCON project funding requests as part of the annual President's budget request. In most instances, Congress appropriates and authorizes in a single budget year MILCON funding for individual projects in a lump sum for the entire project duration, regardless of how long the construction duration. Normally, the money remains available for obligation for five years.

The legal authorities and requirements for projects vary depending on a host of factors, including the size of the project and the circumstances that may require it. Unique authorities exist for national emergencies, for projects supporting contingency operations, and for projects that foster energy resilience and energy conservation.

Several MILCON-related statutes require DOD officials to notify Congress, specifically the "appropriate committees of Congress," about the intent to exercise legal or budgetary authorities. Title 10 U.S.C. §2801 defines such committees to mean the congressional defense committees (i.e., the House and Senate Committees on Armed Services and Appropriations).[7]

---

such portion of a complete and usable facility or improvement as is specifically authorized by law)."

[5] In addition to the definition in Chapter 169 of Title 10, the term military installation under 16 U.S.C. §670(1)(A) means "any land, or interest in land owned by the United States and administered by the Secretary of Defense or the Secretary of a military department, except land under the jurisdiction of the Assistant Secretary of the Army having responsibility for civil works."

[6] 10 U.S.C. §2801.

[7] Reporting requirements, with respect to any project to be carried out by, or for the use of, an intelligence component of the Department of Defense, also include the Permanent Select Committee on Intelligence of the House of Representatives and the Select Committee on Intelligence of the Senate.

Those statutes may also impose a required waiting period before the Secretary concerned can move forward with a project. This waiting period provides Congress an opportunity for oversight, allowing time for Members or committees to ask questions or provide additional direction about individual projects.

## Authority for MILCON Projects

The annual National Defense Authorization Act (NDAA) provides authority for DOD's construction activities.[8] Pursuant to 10 U.S.C. §2802, the Secretary of Defense and the Secretaries of the military departments may carry out such MILCON projects, land acquisitions, and defense access road projects "as are authorized by law."[9] The section goes on to authorize the following activities:

- surveys and site preparation;
- construction, acquisition, conversion, rehabilitation, and installation of facilities;
- acquisition and installation of equipment and appurtenances integral to the project;
- acquisition and installation of supporting facilities (including utilities) incident to the project; and
- planning, supervision, administration, and overhead incident to the project.

In the practical application of these statutes, Congress typically authorizes individual military construction projects, at a line-item level, in the annual NDAA, which originates in the House and Senate Committees on Armed Services (HASC and SASC). Separately, Congress typically appropriates funding for such projects in the annual Military Construction, Veterans

---

[8] The MILCON-unique authorization requirements specified in this section of law are in addition to those established by 10 U.S.C §114, which states that, for specified DOD programs including MILCON, "no funds may be appropriated for any fiscal year ... unless funds therefor have been specifically authorized in law."

[9] 10 U.S.C. §2821 holds a similar requirement for the construction and acquisition of military family housing.

Affairs, and Related Agencies Appropriations Act, which originates in the House and Senate Committees on Appropriations (HAC and SAC).[10]

## Activities Funded Through MILCON Appropriations

The annual Military Construction, Veterans Affairs, and Related Agencies Appropriations Act provides funding (i.e., budgetary authority) for DOD military construction activities. Often referred to as the MILCON-VA appropriations bill, this piece of legislation, in general, appropriates project-specific amounts for major MILCON projects and a variety of other construction and construction-related activities. Components of the legislation include

- *Military Construction.* Includes funding for specific major construction projects for various military components; the legislation typically lists specific dollar amounts for each individual project.[11]
- *Unspecified Minor Military Construction.* Includes funding for smaller construction projects—that is, those estimated to cost no more than the current threshold of $9 million as specified in 10 U.S.C. §2805. Unlike major MILCON projects, which by definition exceed that threshold, unspecified minor construction projects are not subject to specific congressional authorization; therefore, the amounts are appropriated as a lump sum for *unspecified minor military construction*, also sometimes known as UMMC. These funds can be allocated to specific projects at the discretion of the Secretary concerned in accordance with congressional notification requirements and other restrictions.
- *Planning and Design.* Includes funding for certain activities related to planning and design of military construction projects. Prior to authorizing and appropriating funding for a MILCON project,

[10] For more information on authorizations and appropriations, see CRS Report R46497, *Authorizations and the Appropriations Process*, by James V. Saturno; and U.S. Government Accountability Office (GAO), *Principals of Federal Appropriations Law, Fourth Edition, Chapter 2, 2016 Rev.*, GAO-16-464SP, March 10, 2016, pp. 2-54, at https://www.gao.gov/legal/red-book/overview.

[11] MILCON appropriations are usually available for obligation for five fiscal years from the start of the fiscal year in which they are initially appropriated, after which time the appropriation expires (though remains available for certain limited purposes). After an additional five years, any unexpended MILCON funds are canceled and returned to the U.S. Treasury.

Congress generally requires that the project planning and design be sufficiently developed to include a reliable cost estimate and be ready for near-term execution during the current budget year. To meet these requirements, the design phase usually begins before a project is submitted for consideration in the authorization and appropriations process.

- *Family Housing (New Construction and Construction Improvements).* Includes funding for replacement, acquisition, expansion, addition, extension, and alteration of government-owned military family housing.[12]
- *Family Housing (Operations and Maintenance).* Includes funding for routine maintenance, utilities, and the general management of government-owned family housing (including government-owned family housing at overseas locations) as well as debt payment, leasing, minor construction, principal and interest charges, and insurance premiums.
- *Military Privatized Housing.* Congress appropriates funding for the *Family Housing Improvement Fund (FHIF)* and the *Military Unaccompanied Housing Improvement Fund (MUHIF)*, separately.[13] Appropriations to these funds allow the military services to enter into agreements with private housing companies selected in a competitive process to own, maintain, and operate family housing through a long-term lease. DOD calls these activities the Military Housing Privatization Initiative (MHPI).[14]
- *Base Realignment and Closure (BRAC).* Includes funding for functions associated with the implementation of past BRAC rounds (i.e., continuing environmental restoration and caretaker costs) or, if authorized by Congress, any new BRAC round.[15]

---

[12] Construction of government-owned military housing for single service members (often referred to as *Unaccompanied Housing*) is typically funded as major MILCON (e.g., a new barracks). However, military family housing is funded through a separate component of the Military Construction budget that is titled "Family Housing."

[13] See subchapter IV of 10 U.S.C. Ch. 169. Treasury funds for FHIF and MUHIF are established under 10 U.S.C.§2883.

[14] See 10 U.S.C.§2883, "Department of Defense Housing Funds."

[15] Congress continues to provide funding for the implementation of fiscal year (FY) 2005 BRAC decisions. FY2023 appropriations for BRAC totaled $435 million. For more information on BRAC implementation, see the U.S. Government Accountability Office (GAO) report GAO-22-105207, Base Realignment and Closure: DOD Should Provide Congress More Complete and Transparent Information," Sep 28, 2022, at https://www.gao.gov/products/gao- 22-105207.

- *North Atlantic Treaty Organization (NATO) Security Investment Program.* Includes U.S. contributions to the acquisition and construction of common-use military facilities and installations (including international military headquarters), and for related expenses for the collective defense of the North Atlantic Treaty Area.[16]

For a detailed look at the MILCON legislation for FY2024, see CRS Insight IN12238, *FY2024 Military Construction Appropriations: A Summary*, by Andrew Tilghman, and CRS Insight IN12229, *FY2024 NDAA: Military Construction and Housing Authorizations*, by Andrew Tilghman.

## Selected MILCON Authorities

Title 10, Chapter 169, of the U.S. Code establishes several distinct authorities that allow DOD to execute MILCON projects and expend funds for other facilities-related activities.[17] Each authority typically has a monetary threshold or other limitation and these restrictions set the framework for the MILCON budget and authorization process. They are directly tied to congressional oversight of DOD's MILCON-related activities.

When examining notification requirements, monetary thresholds, and other statutory limitations, it is important to note whether the specific authority applies to the Secretary of Defense, or to each Secretary concerned individually.[18]

Most military construction activities are limited to funding that Congress authorizes and appropriates in MILCON accounts. However, there are certain authorities that allow DOD to use funds from Operation and Maintenance (O&M) accounts for military construction activities in certain situations. The services cite "minor construction" in numerous budget justification documents related to O&M accounts. For example, in the FY2025 budget submission, the Army cited "minor construction" as a component of its request for "Base

---

[16] Title 10 U.S.C. §2806. The 29 member states of NATO share the expense of common-use facilities by a formula under which the United States is responsible for about 21% of the annual construction budget of the alliance.

[17] Chapter 169 of Title 10 also includes authorities that support military housing.

[18] As noted above, in prescribing the various MILCON authorities and responsibilities, Chapter 169 regularly uses the term Secretary concerned, referring to the Secretaries of the military departments, or the Secretary of Defense in the case of a defense agency. These terms are defined under 10 U.S.C. §2801.

Operations Support."[19] The Navy cited "minor construction" in its FY2025 justification documents as one of numerous activities carried out with funds from the Sustainment, Restoration and Modernization account.[20] These are examples of requested allocations within the O&M account that DOD or the military services may use for construction projects executed under MILCON-related authorities. Statutes also allow for the use of money from Research, Development, Test and Evaluation (RDT&E) accounts in certain circumstances. Appendix A provides a reference table of selected statutes governing MILCON and repair authorities, lists their limitations, where applicable, and notes instances where O&M or RDT&E funds may be used under MILCON authority.

***Unspecified Minor Construction (10 U.S.C. §2805)***

Title 10, Section 2805, of the *U.S. Code* authorizes the Secretary concerned to use an expedited process for the authorization of *Unspecified Minor Military Construction* projects—currently defined as MILCON projects costing $9 million or less (adjusted annually to reflect the area cost factor).[21] Construction of new military family housing units is specifically prohibited under this section.[22]

---

[19] Department of the Army, Fiscal Year (FY) 2025 Budget Estimates; Operation and Maintenance, Army; see page 191, Budget Activity 01: Operating Forces; Activity Group 13: Land Forces Readiness Support; Detail by Subactivity Group 131: Base Operations Support. At https://www.asafm.army.mil/Portals/72/Documents/BudgetMaterial/ 2025/ Base%20Budget/Operation%20and%20Maintenance/Regular%20Army%20Operation%20 and%20Maintenance%20Volume%201.pdf#page=206.

[20] Department of the Navy, FY 2025 President's Budget Submission, Operation and Maintenance, Navy Budget Activity: Operating Forces, Activity Group: Base Support, Subactivity Group: Sustainment, Restoration and Modernization at https://www.secnav.navy.mil/ fmc/fmb/Documents/25pres/OMN_Book.pdf#page=395.

[21] Through FY2027, the Secretary concerned is required to adjust the dollar limitations specified in 10 U.S.C. §2805 for unspecified minor military construction projects inside the United States to reflect the local area construction cost index for military construction projects; no limitation may exceed $10 million. An area cost factor, also called a construction cost index, is an indicator of the average cost over time of representative goods and services related to construction cost variations. It serves as a reflection of the inflationary or deflationary changes of a specific sector of construction industry. For more information on unspecified minor military construction, see DOD, *Financial Management Regulation* (7000.14-R), Volume 3, Chapter 17, p. 17-15, July 2021, at https://comptroller.defense. gov/Portals/45/documents/fmr/current/03/03_17.pdf.

[22] 10 U.S.C. §2805(e).

Under a temporary provision enacted by the FY2024 NDAA, the commander of the Indo-Pacific Command is authorized to carry out minor military construction projects that cost less than $15 million.[23]

In general, the Secretary concerned may allocate funds for specific projects from the Unspecified Minor Construction accounts, which Congress authorizes in the NDAA and funds in the annual appropriations act for this purpose. Examples of Unspecified Minor Military Construction might include the construction of new security gates at the entrance to an installation, new vehicle maintenance sheds or additions to existing buildings. Under this authority, any project costing more than $750,000 requires advance approval by the Secretary concerned. In instances where the estimated total cost of a project is $4 million or less, the funds may be drawn from MILCON accounts or O&M accounts such as "Base Support Operations," "Sustainment, Restoration and Modernization," or other accounts.

For Unspecified Minor Military Construction projects exceeding $4 million, 10 U.S.C. §2805 requires the Secretary concerned to submit a notification to the *appropriate committees of Congress*.[24] This notification is to include a justification for the project and the estimated cost. Following a 14-day waiting period, and pending resolution of any questions or concerns from Congress, the project can then be executed.

### *Laboratory Modernization and Revitalization (10 U.S.C. §2805(d) and §2805(g))*

The Secretary of Defense has authority that provides flexibility for funding military construction projects that support the DOD's laboratories and research facilities. The James M. Inhofe National Defense Authorization Act for Fiscal Year 2023 authorized and funded the "Defense Laboratory Modernization Program," which grants authority for the Secretary of Defense to use funds from RDT&E accounts to carry out certain military construction

---

[23] P.L. 118-31, §2810. The temporary provision expires March 31, 2029.

[24] 10 U.S.C. §2801 defines the term *appropriate committees of Congress* as the "congressional defense committees and, with respect to any project to be carried out by, or for the use of, an intelligence component of the Department of Defense, the Permanent Select Committee on Intelligence of the House of Representatives and the Select Committee on Intelligence of the Senate." 10 U.S.C. §101 defines the term *congressional defense committees* as "the Committee on Armed Services and the Committee on Appropriations of the Senate; and the Committee on Armed Services and the Committee on Appropriations of the House of Representatives."

projects supporting technology research that are authorized in law.[25] The statute requires the Defense Secretary to submit project descriptions, justifications and cost estimates with the annual President's budget request for inclusion in the annual MILCON authorization and appropriations acts. While there is no specific limit on funding for individual projects carried out under this authority, the statute does impose a maximum amount of $150 million for RDT&E funds that may be obligated in any fiscal year for military construction projects.[26] In addition, 10 U.S.C. §2805 provides authority for the use of O&M money to fund certain unspecified minor construction projects intended to enhance DOD laboratories that conduct research, engineering, testing and evaluation activities.[27]

***Architectural, Engineering, and Design Services (10 U.S.C. §2807)***

Title 10, Section 2807, of the *U.S. Code* allows the Secretary concerned to obtain architectural and engineering services, and to initiate construction design, in connection with MILCON projects that are not otherwise authorized by law.[28] If the required services are estimated to exceed $1 million, congressional notification is required and the Secretary concerned must wait 14 days before obligating funding for such purposes. Funding under this authority allows for the DOD to develop a project's planning documents so that the project has a reliable cost estimate to provide to Congress when it is submitted for a funding request during the current year budget cycle and so the project is potentially ready for execution during the fiscal year of its funding.

Initial activities of the MILCON planning process—activities such as installation master planning, project identification, and initial environmental reviews—may not be paid for with MILCON funds. Instead, these activities

[25] See 10 U.S.C. §2805(g). Also see FY2023 NDAA (P.L. 117-263). The use of RDT&E money for military construction projects under the Sect. 2805 authority is limited to projects at the statutorily defined DOD Science and Technology Reinvention Laboratories; projects for a DOD federally funded Research and Development Centers that functions primarily as a research laboratory, and projects at DOD facilities in support of a technology development program that is consistent with the fielding of offset technologies [defined as the Advanced Sensor Applications Program].

[26] See 10 U.S.C. §2805(g). Also see FY2023 NDAA (P.L. 117-263)

[27] 10 U.S.C. §2805 defines the term *laboratory* to include "a research, engineering, and development center" and "a test and evaluation activity."

[28] The statute allows for the secretary concerned to obtain architectural and engineering services, and to initiate construction design for projects related to family housing, and projects undertaken in connection with 10 U.S.C. §2854, "Restoration or replacement of damaged or destroyed facilities."

are typically funded with appropriations from O&M accounts, Research, Development, Testing and Evaluation (RDT&E) or Procurement accounts. Those activities—often called "advance planning"—can occur during the initial analysis of an emerging military requirement and before the service determines the need for a major MILCON project to meet that requirement. Only after the project development gets underway, and planners identify a clear requirement for a new construction project, are MILCON funds used to pay for certain essential parts of the design process, such as engineering and architectural documents for the bidding or construction phase. Funding for these activities are typically authorized and appropriated under a category within the MILCON budget known as "Planning and Design."

### *Repair of Facilities (10 U.S.C. §2811)*

Title 10, Section 2811, of the *U.S. Code* allows the Secretary concerned to carry out certain "repair" projects on facilities using O&M appropriations.[29] The statute defines a repair project as a project "(1) to restore a real property facility, system, or component to such a condition that it may effectively be used for its designated functional purpose; or (2) to convert a real property facility, system, or component to a new functional purpose without increasing its external dimensions." There are no budgetary limitations or caps on this authority, but repair projects costing more than $7.5 million must be approved in advance by the Secretary concerned, and Congress must be notified of the decision to carry out the project. The notification must include a cost estimate and a justification for the project. If the cost estimate exceeds 75% of the estimated cost of a MILCON project to replace the facility, the statute also requires the congressional notification to include an explanation as to why replacement of the facility is not in the best interest of the government. This authority prohibits the construction of new facilities or additions to existing facilities.

The statute for "Repair of Facilities" is broadly written and can be used for a range of projects that do not include expanding the square footage of a structure. For example, converting a warehouse into office space might be characterized as a repair if it meets the statutory requirements. This statute may also provide for the upgrading of HVAC systems or utility systems. The authority to use O&M funds for repairs may provide budget flexibility for the military departments, allowing for the execution of improvements and fixes

[29] The O&M sub-accounts associated with this type of maintenance work are generally referred to as Facilities Sustainment, Restoration, and Modernization, or FSRM.

on a more rapid timeline compared to the traditional MILCON process. The approval process for repair projects is less complex and time consuming, in part because repair projects do not necessarily have to directly compete for priority funding with military construction projects across the entire service.

Examples of exercising this authority include a Navy project using $7.9 million in O&M money for projects that included converting a barracks at Marine Corps Base Camp Pendleton into a Consolidated Training Facility for the 1st Marine Division schools.[30] Another Navy example was a $5.3 million project at the Navy base in Rota, Spain, which converted a vacant administration facility into a command-and-control space supporting forward deployed Naval forces.[31]

### *Military Construction Projects for Energy Resilience, Energy Security, and Energy Conservation (10 U.S.C. §2914)*

The Energy Resilience and Conservation Investment Program (ERCIP) is funded as a subset of the Military Construction, Defense-Wide account. The ERCIP is intended to fund projects for the military services and other components that improve energy resilience, contribute to mission assurance, save energy, and reduce DOD's energy costs.[32] ERCIP accomplishes this through construction of high-efficiency energy systems and technologies or through modernizing existing energy systems.

ERCIP activities are authorized under Title 10, Section 2914, of the U.S. Code, which allows the Secretary of Defense to improve the energy resilience, energy security, and energy conservation of its fixed installations. Energy resilience is defined as "the ability to avoid, prepare for, minimize, adapt to, and recover from anticipated and unanticipated energy disruptions in order to ensure energy availability and reliability sufficient to provide for mission assurance and readiness, including mission essential operations related to readiness, and to execute or rapidly reestablish mission essential requirements."[33]

---

[30] Department of the Navy Installation Energy Resilience Strategy, p. 24, February 2020, at https://www.secnav.navy.mil/eie/Documents/DON-Installation-Energy-Resilience-Strategy.pdf.

[31] Ibid.

[32] Mission assurance is a term to describe the ability of a unit or facility to continue performing and achieving military missions despite any unexpected challenge that may arise, such as a deliberate attack or natural disaster. The statutory definition of energy resilience is included in 10 U.S.C. §101.

[33] Statutory definition of energy resilience is included in 10 U.S.C. §101.

The ERCIP, authorized by Congress in Section 2851 of the John Warner National Defense Authorization Act for Fiscal Year 2007 (P.L. 109-364), has focused on projects that support energy resilience for critical mission requirements to help ensure that military installations can continue to maintain operational readiness in the event of a crisis, such as a natural disaster or cyberattack that damages the traditional energy grid.[34] The department has sought to prioritize energy resilience improvement projects on installations listed on the DOD Mission Assurance Priority List, which is a list of installations across the military that are considered most critical to readiness.[35]

Total appropriations provided for the ERCIP program for FY2024 were $634 million.[36] Examples of ERCIP projects funded in the FY2024 MILCON-VA Act included $57 million for a new microgrid with backup system at Vandenberg Space Force Base in California.[37] Another example was authorization of $25 million for a microgrid and battery storage system for F.E. Warren Air Force Base in Wyoming.[38]

ERCIP also funds water resilience projects that improve or enable access to water in support of mission functions; however, ERCIP does not fund projects that are solely for water compliance, meaning a project whose sole

---

[34] Statement of Mr. Paul Cramer, performing the duties of Assistant Secretary of Defense (Energy, Installations & Environment), testimony before the Senate Committee on Appropriations Subcommittee on Military Construction and Veterans Affairs, May 18, 2022, at https://www.appropriations.senate.gov/imo/media/doc/ASD(EIE)%20Cramer_OSD_%20SAC-M%20MILCON%20and%20Family%20Housing%20Hearing%20Statement%20(5.18.22).pdf.

[35] DOD, Office of the Under Secretary of Defense for Sustainment, Fiscal Year 2024 Energy Resilience and Conservation Investment Program Guidance from the Deputy Assistant Secretary of Defense (Environment and Energy Resilience), October 15, 2021, at https://www.acq.osd.mil/eie/Downloads/IE/FY2024%20ERCIP%20Program%20Guidance.pdf.

[36] See Joint Explanatory Statement accompanying P.L. 118-42 P.L. 118-42, Division A, Military Construction, Veterans Affairs, and Related Agencies Act, 2024, page 56, at https://www.govinfo.gov/content/pkg/CPRT-118HPRT55007/pdf/CPRT-118HPRT55007.pdf#page=62.

[37] See P.L. 118-31, PDF page 775, at https://www.congress.gov/118/plaws/publ31/PLAW-118publ31.pdf. Also see DOD Comptroller, FY 2024 Energy Resilience and Conservation Investment Program (ERCIP) Project List by State/Country, https://comptroller.defense.gov/Portals/45/Documents/defbudget/fy2024/budget_justification/pdfs/07_Military_Construction/15-Energy_Resilience_and_Conservation_Investment_Program.pdf#page=10.

[38] See P.L. 118-31, PDF page 778, at https://www.congress.gov/118/plaws/publ31/PLAW-118publ31.pdf#page=778. Also see DOD Comptroller, FY 2024 Energy Resilience and Conservation Investment Program (ERCIP) Project List by State/Country, at https://comptroller.defense.gov/Portals/45/Documents/defbudget/fy2024/budget_justification/pdfs/07_Military_Constr uction/15-Energy_Resilience_and_ Conservation_Investment_ Program.pdf#page=37.

objective is to enable installations to meet environmental and permitting standards.[39]

The law allows for the Secretary concerned to use MILCON funding for these projects. It also allows for the use of O&M funding for energy conservation and resilience projects if the Secretary concerned notifies the congressional defense committees about the intent to do so. Such notifications require detailed information that includes: the source of those O&M funds to be used; the reasons why the project needs to be prioritized; and, in the case of projects designed specifically for energy conservation, a savings-to-investment ratio that would result from the improved energy efficiencies. The Secretary concerned is limited to a maximum obligation of $100 million in O&M funding under this authority for any fiscal year.[40] Projects can be carried out only after a seven-day waiting period following notification of the appropriate congressional defense committees.

***Projects Supporting Installation Resilience (10 U.S.C. §2815)***

The Secretary of Defense has certain authorities to carry out projects that the Secretary determines would enhance military installation resilience, energy security, and energy conservation or address known vulnerabilities of fixed installations.[41] Under 10 U.S.C. §2815, the Secretary may carry out projects *at locations outside of a military installation or facility* if the project would "preserve or enhance" the resilience of a military installation or community infrastructure that the Secretary determines is "necessary to maintain, improve

---

[39] According to DOD policy guidance, proposed projects that are solely for water compliance (i.e., a project whose sole objective is to enable installations to meet environmental and permitting standards) will not be considered or selected for submission to Congress. DOD will consider water resilience and water security projects for submission when they contribute to overall energy resilience. See DOD, Office of the Under Secretary of Defense for Sustainment, Fiscal Year 2024 Energy Resilience and Conservation Investment Program Guidance from the Deputy Assistant Secretary of Defense (Environment and Energy Resilience), Section 2.4, October 15, 2021, at https://www.acq.osd.mil/eie/Downloads/IE/FY2024%20ERCIP%20Program%20Guidance.pdf#page=8.

[40] 10 U.S.C. §2914(c)(3).

[41] 10 U.S.C. §101(e)(8) defines the term *military installation resilience* as "the capability of a military installation to avoid, prepare for, minimize the effect of, adapt to, and recover from extreme weather events, or from anticipated or unanticipated changes in environmental conditions, that do, or have the potential to, adversely affect the military installation or essential transportation, logistical, or other necessary resources outside of the military installation that are necessary in order to maintain, improve, or rapidly reestablish installation mission assurance and mission-essential functions." For more information, see CRS Insight IN11566, *Military Installation Resilience: What Does It Mean?*

or rapidly establish installation mission assurance and mission-essential functions." The Secretary may also carry out stormwater management projects if those projects support the resilience of a military installation, a defense access road, or "other essential civilian infrastructure." Such stormwater management projects may also be carried out for the purpose of "protecting nearby waterways and stormwater-stressed ecosystems."

For projects in support of military installation resilience, the Secretary may use funding from O&M accounts if the Secretary notifies Congress about the estimated cost of the project, the source of funds for the project, and a certification that waiting to seek funding in the next budget cycle under the next Military Construction Authorization Act "would be inconsistent with national security or the protection of health, safety or environmental quality." Such projects are subject to a seven-day waiting period following congressional notification (and pending resolution of any concerns articulated by Congress). The total maximum amount of O&M funding that can be obligated under Section 2815 cannot exceed $125 million in any fiscal year.[42]

### *Authority for Certain Construction Projects in Friendly Foreign Countries (10 U.S.C.§2817)*

Under Title 10, Section 2817, the Secretary of Defense may carry out certain construction projects outside of a military installation in a friendly foreign country if the geographical combatant commander determines that the construction project is "necessary to support vital United States military requirements at an air port of debarkation, sea port of debarkation, or rail or other logistics support location."[43] The statute requires the Secretary to make a determination that deferral of the construction project for inclusion in the next NDAA is inconsistent with military requirements and other national security or national interests of the United States. The Secretary is authorized the use of O&M funds for projects carried out under this statute.

Individual projects carried out under this authority are limited to a total cost of $15 million, and the statute limits the Secretary from obligating more than $200 million in any fiscal year under the authority. The Secretary of Defense is required to notify the congressional defense committees about projects carried out under this authority and such projects are subject to a 30-day waiting period after congressional notification.

---

[42] 10 U.S.C. §2815(e)(3).
[43] 10 U.S.C. §2817.

### *Emergency Authorities*[44]

The following DOD authorities are intended to be exercised in the event of a national emergency, declaration of war, or other acute circumstance in which the Secretary concerned determines a MILCON project is immediately required to meet national security needs, or in which waiting for the next MILCON authorization act would be inconsistent with national security or national interest.[45] These authorities allow for funding and executing MILCON projects outside the traditional PPBE process in certain circumstances.

### *Emergency Construction (10 U.S.C. §2803)*

Under Title 10, Section 2803, of the *U.S. Code*, the Secretary concerned may carry out a MILCON project not otherwise authorized in law if he or she determines that (1) it is vital to national security or to the protection of health, safety, or the quality of the environment, and (2) the project is of such urgency that it cannot wait for the next military construction authorization act (i.e., the NDAA).[46] When the Secretary concerned initiates a project under this section, the statute requires the Secretary concerned to notify the appropriate congressional committees in writing. The congressional notification must include the justification for the project, the estimated cost of the project, the reason for invoking Section 2803, and the source of the funds to be used. Accordingly, when requesting approval for an emergency construction project, DOD submits a notification and request for the reprogramming of MILCON funds in accordance with this statute.[47] There is a limit of $50 million that can be obligated under this authority during any given fiscal year. Additionally, a project carried out under this authority must be funded within the total amount of available (unobligated) MILCON appropriations that have

---

[44] These authorities are generally not applicable to the planning, programming, and budgeting process.

[45] 10 U.S.C. §2804.

[46] There is no provision in Title 10 that explicitly grants this authority to the reserve components of the armed forces (as they are defined in Chapter 1003 of Title 10), and the DOD does not consider emergency construction authority as applicable to the reserve components. The reserve component's real property acquisition authorities are provided in Chapter 1803 of Title 10. See DOD, *Financial Management Regulation* (7000.14-R), Volume 3, Chapter 17, p. 17-19, July 2021, at https://comptroller.defense.gov/Portals/45/documents/fmr/current/03/03_17.pdf.

[47] For more information on DOD regulations pertaining to emergency construction authority, see DOD, *Financial Management Regulation* (7000.14-R), Volume 3, Chapter 17, pp. 17-18, July 2021, at https://comptroller.defense.gov/Portals/45/documents/fmr/current/03/03_17.pdf.

not been obligated.[48] There is also a mandatory waiting period of five days, beginning on the date the Secretary concerned notifies the appropriate committees of Congress, before the emergency construction project may begin.

### *Contingency Construction (10 U.S.C. §2804)*

Section 2804 of Title 10 of the U.S. Code permits the Secretary of Defense to use MILCON funds *appropriated specifically for use under this authority* to carry out a MILCON project without prior congressional authorization, if the Secretary determines that waiting to include it in the next MILCON authorization act would be "inconsistent with national security or national interest."

The authority under Section 2804 is rarely invoked and is generally reserved for projects requested by combatant commands supporting multi-service requirements.[49] Projects must be carried out using funds specifically appropriated for this authority (which historically have been appropriated as a line item under the Military Construction, Defense-Wide account.)[50] The DOD has a unique process for approving and prioritizing requests under this authority. Combatant commanders verify requests and forward them through the Chairman of the Joint Chiefs of Staff (CJCS), who is responsible for assigning priority among competing requests.[51] The CJCS then forwards them to the Office of the Secretary of Defense (OSD). OSD may also receive requests from the individual military departments and is responsible for coordinating the congressional notifications.[52]

---

[48] The Military Construction, Veterans Affairs, and Related Agencies Appropriations Act, 2023, Division J, Sec. 116, states: "...any funds made available to a military department or defense agency for the construction of military projects may be obligated for a military construction project or contract, or for any portion of such a project or contract, time before the end of the fourth fiscal year after the fiscal year for which funds for such project were made available." Found at https://www.congress.gov/117/bills/hr2617/BILLS-117hr2617enr.pdf#page=486.

[49] See GAO, *Defense Infrastructure: Actions Needed to Enhance Oversight of Construction Projects Supporting Military Contingency Operations*, GAO 16-406, September 8, 2016, at https://www.gao.gov/products/gao-16-406.

[50] For example, the conference report (H.Rept. 112-331) accompanying the Consolidated Appropriations Act, 2012 (P.L. 112-74) listed $10 million for contingency construction, within the Military Construction, Defense-Wide account. See H.Rept. 112-331, p. 1283. Versions of the legislation since 2012 have not included appropriations for contingency construction under 10 U.S.C. §2804.

[51] See GAO, *Defense Infrastructure: Actions Needed to Enhance Oversight of Construction Projects Supporting Military Contingency Operations*, GAO-16-406, September 8, 2016, at https://www.gao.gov/products/gao-16-406.

[52] Ibid.

Section 2804 requires the Secretary to submit a report in writing to the appropriate congressional committees justifying the project and the invocation of this authority and including cost estimate. Further, although DOD may initiate a MILCON project under this authority without specific congressional authorization, actual construction may not begin prior to the receipt of appropriate DOD and congressional approval.[53]

### *Construction Authority in the Event of a Declaration of War or National Emergency (10 U.S.C. §2808)*

In the event of a declaration of war or declaration of a national emergency by the President *that requires the use of the armed forces*, 10 U.S.C. §2808 permits the Secretary of Defense to initiate, or to authorize the Secretaries of the military departments to carry out, MILCON projects not yet authorized in law.[54] Any unobligated MILCON or family housing appropriations may be used for construction projects that are "necessary to support such use of the armed forces." According to the statute, this authority may be exercised "without regard to any other provision of law." The Secretary of Defense must report to Congress the decision to use this authority, along with the estimated costs of the projects, including any associated real estate transactions. The authority terminates at the end of the war or national emergency.

In the William M. (Mac) Thornberry National Defense Authorization Act for Fiscal Year 2021 (P.L. 116-283), Congress imposed caps on the amount of total funding that the Secretary may initiate under 10 U.S.C. §2808 and also imposed additional congressional notification requirements.[55] Currently, the total cost of all military construction projects undertaken using the authority during the national emergency may not exceed $500 million. For funding used for projects only within the United States, the total cost may not exceed $100 million.

The statute requires the Secretary of Defense to provide a detailed notification that includes the cost of the projects planned under this authority and the impact that would have on previously planned MILCON projects that

---

[53] For more information on Contingency Construction, see DOD, *Financial Management Regulation* (7000.14-R), Volume 3, Chapter 17, Section 3.3.3, pp. 17-19 to 17-20, July 2021, at https://comptroller.defense.gov/Portals/45/documents/fmr/current/03/03_17.pdf.

[54] 10 U.S.C. §2808 states that a declaration of a national emergency must be in accordance with the National Emergencies Act of 1976 (50 U.S.C. §1601 et seq.). The National Emergencies Act does not define the term "national emergency." See CRS Legal Sidebar LSB10267, *Definition of National Emergency under the National Emergencies Act*, by Jennifer K. Elsea.

[55] P.L. 116-283, §2801; 134 Stat. 4317-4319.

may be delayed or canceled to fund the emergency project requested.[56] The Secretary may carry out the project after a five-day waiting period.

For more information on use of 10 U.S.C. §2808 authority, see CRS Insight IN11017, *Military Construction Funding in the Event of a National Emergency*; and CRS Report R45937, *Military Funding for Southwest Border Barriers*. Also see CRS Report R47979, *DHS Border Barrier Funding Developments: FY2021-FY2024*, by William L. Painter.

### *Restoration or Replacement of Damaged or Destroyed Facilities (10 U.S.C.§2854)*

Title 10, Section 2854, of the *U.S. Code* provides unique authorities to fund the restoration or replacement of damaged or destroyed facilities. The Secretary concerned may authorize the use of MILCON funding for projects under this authority subject to notification of the appropriate congressional committees. The notification must include the project's current cost estimate, a justification for carrying out the project under this authority and notification of the source of the funds intended for use. Projects using MILCON funding under this authority are subject to a 14- day waiting period following congressional notification.

The Secretary concerned also may authorize the use of O&M funding for MILCON projects under this authority if the damage at issue was the result of a natural disaster or a terrorism incident. For example, after Hurricane Michael inflicted damage on Tyndall Air Force Base in Florida in 2018, the Air Force invoked this authority for the first time to expedite some new construction.[57] To exercise this authority, the Secretary concerned must notify the appropriate congressional committees and provide information that includes: the project's current cost estimate; certification that waiting for authorization under the next Military Construction Authorization Act would be inconsistent with national security or the protection of health, safety, or environmental quality; and, in the case of damage to a facility rather than destruction, a certification that the replacement project is more cost-effective than repair or restoration. Use of O&M funds under this authority are subject to a seven-day waiting period following congressional notification. The statute also limits the Secretary

---

[56] See §2808(e) Notification Requirement.

[57] Defense Department, "Tyndall Rebuild taps into unique funding for New Fire Stations," Oct. 17, 2022, at https://www.dvidshub.net/news/431910/tyndall-rebuild-taps-into-unique-funding-new-fire-stations.

concerned to a maximum of $100 million in O&M funding under this authority for any fiscal year.

## Planning, Programming, and Budgeting[58]

Initial ideas for a potential MILCON project can come from a variety of sources. The need for a particular construction project may be part of a top-down initiative (e.g., the creation of new missions for a given military installation or the movement of units or offices between installations) or it may be the result of modernization or replacement of existing facilities determined at the local level (i.e., by installation commanders and other local stakeholders).

Some planning and design work must begin before DOD will prioritize a project for a funding request in the PPBE process. Congress requires requests for authorization and appropriation to include only those projects that have a reliable cost estimate and projects that can realistically be executed during the fiscal year authorized, so requests must show documented progress on planning and design. Insufficient planning and design development can prevent a project from receiving approval for the current budget year.

The progress of a particular project's planning and design is considered to be an indicator of the accuracy of its cost estimate. Senior military leaders and the White House's Office of Management and Budget (OMB) may be reluctant to put forth a particular project for the current budget year if the planning and design work remains in an early phase, because incomplete design work increases the risk that a projected cost will increase and result in the need for reprogramming or additional funding authorization and appropriation. Projects and their cost estimates are rated in accordance with a standardized cost estimating classification system.[59] This rating, on a scale of

---

[58] This section intentionally condenses the DOD's lengthy and complex process for MILCON design and planning, programming, and budgeting. An expanded and more detailed description of the MILCON process can be found within the report, RAND, *Obtaining Life-Cycle Cost-Effective Facilities in the Department of Defense*, 2013, at https://www.rand.org/pubs/research_reports/RR169.html. For more information on the PPBE process, see CRS In Focus IF10429, *Defense Primer: Planning, Programming, Budgeting, and Execution (PPBE) Process*, by Brendan W. McGarry and CRS Report R47178, *DOD Planning, Programming, Budgeting, and Execution (PPBE): Overview and Selected Issues for Congress*, by Brendan W. McGarry.

[59] The Cost Estimating Classification System is a commonly accepted standard across the construction industry and is set by the American Association of Cost Engineers (AACE). For additional information, see AACE International Recommended Practice No. 10S-90,

one to five, offers a metric for the maturity of a project's planning and design work. (For example, a Class 5 estimate indicates the most limited planning and design development, and is therefore more likely to see a change in the cost estimate as the project progresses. Conversely, a Class 1 estimate reflects a fuller project definition, with design and engineering work that is near completion, and likely reflects a more accurate estimate of the final construction costs.) OSD does not have a policy requirement for projects to meet a certain class rating for inclusion in an annual budget. However, a rating that reflects more advanced planning and design work can weigh in favor of a project in the final determination for whether a MILCON project will be included in the President's budget request.[60]

## DD Form 1391

Central to the planning-programming-and-budgeting-process for MILCON projects is the Defense Department (DD) Form 1391, a key document that will accompany the individual project through the entire budget approval and legislative process. (See Appendix B for an example DD Form 1391). Development of the DD Form 1391 for a particular project starts at the early planning stage and continues as work advances on design, architectural documents, and engineering documents. The form's main components include:

- a cost estimate;
- a description of the proposed construction;
- the specific military requirement for the project (e.g., a change in mission or deterioration of facilities already in use);
- the current facility situation at the installation; and
- the impact on military missions if the project is not approved, and any other needed justification information.[61]

---

Cost Engineering Terminology," p. 36, at https://web.aacei.org/docs/default-source/rps/10s-90.pdf?sfvrsn=70.

[60] OMB policy requires that requests for construction funding require that planning and design work be sufficiently developed to "ensure that a contract for construction could be awarded during the budget year." OMB Circular No. A- 11, Preparation, Submission, and Execution of the Budget, July 2024, at https://www.whitehouse.gov/wp-content/uploads/2018/06/a11.pdf#page=119.

[61] DD Form 1391 is associated with several other required DOD forms and their corresponding processes that help to populate, or are informed by, the document. For example, the ENG

Note in Appendix B that the DD Form 1391 requires supplemental information that includes the status of the project's design work. The forms must specifically state the date that the project reached the status of 35 percent design completion, the percent completion at the time of submission, and the date when the design is scheduled to be 100 percent complete.

## Advance Planning

The planning and design process for a military construction project includes two distinct phases. The first phase involves advance planning, which examines the military requirements for a potential project and includes potential site analysis. As part of the initial advance planning phase (see *Figure 1*), installation engineers and other public works staff evaluate the proposed *need* for new or improved facilities—such as training classrooms, firing ranges, barracks, or housing—and compare that to the facilities that already exist.[62] During the initial consideration of the new requirement, military officials may consider alternatives to the MILCON process, such as renovating existing infrastructure or leasing a facility. The installation's engineering team weighs the costs and time required to meet the military requirement by rehabilitating or altering existing structures on site compared to the cost and time required to execute new construction.[63] Engineers and other members of the installation's planning team then make a preliminary determination whether renovation of an existing facility, new construction, or leasing is more appropriate to satisfy the military need[64] This initial phase is funded using money from the military's O&M accounts. In some cases, this

---

FORM 3086 is a tool that cost engineers use to estimate project costs for the DD Form 1391.

[62] Installation engineers are the principal officers responsible for providing authoritative analysis and strategic program recommendations for MILCON needs and priorities. Installation engineers may be referred to in various titles according to each military service: the Post Engineer in the Army, Base Civil Engineer (BCE) in the Air Force and Air National Guard, the Resident/Regional Officer in Charge of Construction (ROICC) in the Navy and Marine Corps, and the Construction and Facilities Management Officer (CFMO) in the Army National Guard.

[63] Rehabilitation, renovation, or alternation of existing structures may be potentially funded with O&M money using a less time-consuming budget approval process. For example, the Repair of Facilities (10 U.S.C. §2811) authority can fund certain types of construction projects without requiring line item-level budget authority and authorization.

[64] Constantine Samaras, et al., *Development, Construction, and Operating Process and Barriers to Life-Cycle Cost-Effectiveness*, RAND Corporation, 2013, Ch. 2, p. 7, at https://www.rand.org/pubs/research_reports/RR169.html.

work is carried out by existing staff. In some instances, a service may specifically request O&M funding for this purpose. For example, in justifying its FY2023 budget request, the Navy requested $214.8 million for a budget item (technically known as a budget sub-activity group, or SAG) called Air Operations and Safety Support. This line finances several operations, including Aviation Facilities and Landing Aids, which supports the "development/revision of aviation facilities planning and design criteria."[65]

The advance planning phase, funded through O&M accounts, may include:

- developing and documenting the military requirements underpinning the proposed military construction project;
- developing or revising an installation's master plan to ensure the proposed project comports with the base's long-term needs;
- conducting site studies to identify a location for a proposed project;
- preparing engineering analysis and development of technical design parameters; and
- preparing environmental impact statements or other requirements under the National Environmental Policy Act (NEPA; P.L. 91-190);

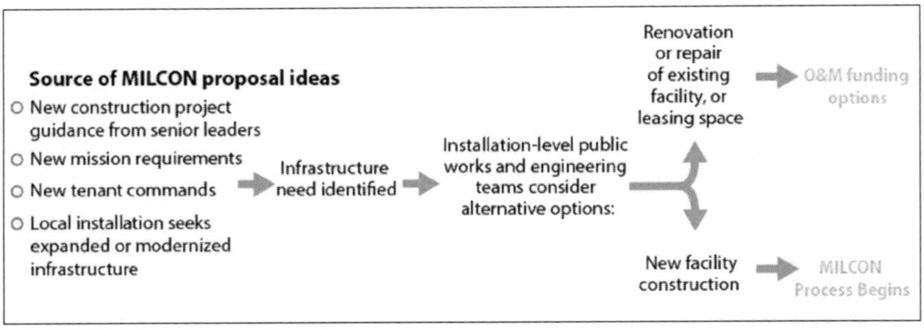


Source: CRS analysis of Department of Defense documents. See for example, DOD, Financial Management Regulation (7000.14-R), Volume 3, Chapter 17, page 17-25, Also see Air Force Instruction 32-1021, February 25, 2016, page 10.

**Figure 1.** Advance Planning for Military Facilities. Preliminary Analysis will determine whether a MILCON project is needed.

---

[65] For more information, see DOD, Department of the Navy, Fiscal Year (FY) 2023 Budget Estimates, Justification of Estimates, Operation and Maintenance, Navy, April 2022, p. 1A4A, available at https://www.secnav.navy.mil/fmc/fmb/Documents/23pres/OMN_Book.pdf#page=79.

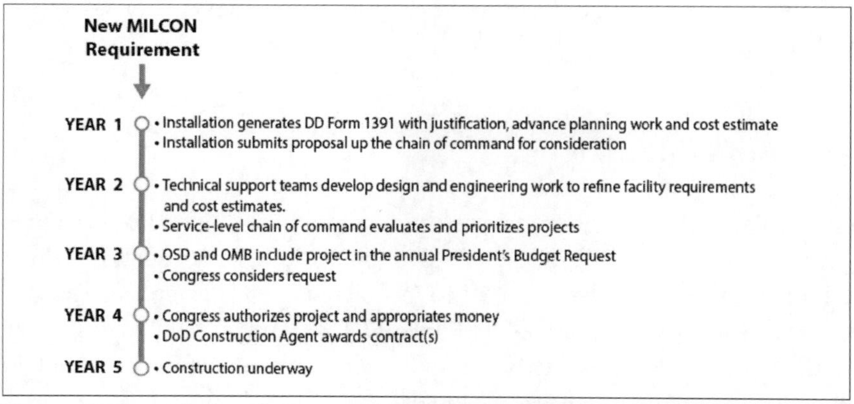


Source: CRS analysis of Department of Defense documents. See for example Department of Defense Directive 4270.5, "Military Construction," updated Aug. 13, 2018, at https://www.esd.whs.mil/Portals/54/Documents/DD/issuances/dodd/427005p.pdf. Also see Commander, Naval Facilities Engineering Systems Command, NAVFAC Instruction 7045.1, "Proper Use of Military Construction Funds."

Notes: Timelines for specific projects will vary depending on their complexity to develop and construct as well as their prioritization by decisionmakers at DOD, White House and Congress.

**Figure 2.** Notional Timeline for Military Construction Projects. The MILCON process begins with the determination that an emerging military requirement warrants a new construction project.

This advance planning process is led by the uniformed installation commander and the installation's public works department, which may consist of engineers, community planners, environmental specialists, and other public works professionals. When an individual installation has competing prospective MILCON projects, the prioritization at the installation level generally begins with meetings between the installation's engineers and representatives of all major resident organizations and other installation tenants. This group is sometimes referred to as an Installation Facilities Board. The result of these meetings—a prioritized list of construction needs—is presented to the installation commander, who then accepts or adjusts the priorities to create a final list of proposed projects for service-level review.

If and when the installation commander determines that a new construction project is necessary, and requires funding through the MILCON process, the team creates a DD Form 1391. From that point of initial determination that a new MILCON project is needed, the formal process can take about five years until the project is approved and construction completed (see Figure 2).

## Design and Engineering

As a project progresses, the advance planning (funded by O&M accounts) transitions to the design phase (funded by MILCON accounts).[66] Funds authorized and appropriated for MILCON "planning and design" may support continuing development of the DD Form 1391, as well as detailed design and construction deliverables, such as request for proposal (RFP) documents or construction drawings and specifications.[67]

The department's annual defense budget typically includes lump-sum appropriations for planning and design that are budgeted for specific components, such as a service branch, a combatant command, or a defense agency.[68] Such funding generally is not tied to any specific project; rather, it is intended to be used to fund progress on planning and design efforts on projects that have not yet been fully funded and authorized by law, or design for projects that remain in the development and prioritization phase.[69] The design phase aims to finalize a reliable cost estimate for Congress to consider

---

[66] Depending on the size and cost of the project, moving into the design phase—and receiving MILCON Planning and Design money—typically requires approval from the military service leadership, secretary's office, or congressional notification. For example, in the Army, authorization to use MILCON Planning and Design funds is based on Army headquarters issuance of a "design code" – See Army Regulation 420-1, page 4-36d(1) " (1) Army planning and design funds will be used for USACE project design activities after issuance of a design code." https://armypubs.army.mil/epubs/DR_pubs/DR_a/pdf/web/ARN15517_R420_1_admin_FINAL.pdf#page=196. Also see the DOD's Financial Management Regulation notes that the secretary concerned has authority to allocate Planning and Design funding as long as the design costs do not exceed the threshold for reporting (see https://comptroller.defense.gov/Portals/45/documents/fmr/archive/03arch/03_17_Dec96.pdf.)

[67] As noted previously, funding for additions or alterations to existing buildings and the construction of new buildings, where costs exceed $4 million, is generally provided through annual MILCON appropriations. Facility repair at a military base is generally supported through the O&M accounts in the annual defense appropriation bill. The O&M sub-accounts associated with this type of maintenance work are generally referred to as Facilities Sustainment, Restoration, and Modernization, or FSRM.

[68] To avoid confusion between O&M funds used for advance planning and the MILCON "planning and design" funds used for design, some DOD officials describe the money appropriated under 10 U.S.C. § 2807 as "MILCON-D." As an example, see NAVFACINST 7045.1, "Proper Use of Military Construction Funds," Dec. 6 2020.

[69] 10 U.S.C. §2807 requires congressional notification if design services are estimated to exceed $1 million.

in the authorization and appropriations process and also to achieve a level of detailed design that will enable the DOD to award a construction contract (or contracts) for the project to potentially begin construction early in the fiscal year of its authorization. The design phase may include

- Site surveys and mapping;
- Detailed architectural drawings and engineering plans; and
- Pre-construction contract award activities, including the preparation of bid documents and liaison with prospective bidders and construction teams.

## DOD Vetting and Prioritization

While most MILCON projects involve initial input from the local installation commander, from there the requests follow different paths depending on the DOD organization responsible for managing a given military installation. Figure 3 depicts these paths. In short, the service branches internally prioritize their MILCON projects and send a final list of top priorities to OSD as part of the annual defense budgeting process. A key part of this process is ensuring that the individual projects have sufficient documentation of planning and design to ensure the reliability of the cost estimate and viability of near-term execution.

When developing MILCON plans, the military services typically work several years in advance of the current budget cycle. (See Figure 4.) While finalizing a list of MILCON projects for inclusion in the upcoming President's budget request, the services are regularly developing and revising the Future Years Defense Program (FYDP), which summarizes the military's five-year plan for the forces, resources, and programs associated with DOD operations.[70] That involves making preliminary decisions about likely

[70] 10 U.S.C. §221 requires that the Secretary of Defense submit, at about the same time as the President's annual budget request, a "future-years defense program ... reflecting the estimated expenditures and proposed appropriations included in that budget. Any such future-years defense program shall cover the fiscal year with respect to which the budget is submitted and at least the four succeeding fiscal years." DOD develops its FYDP according to policies and procedures outlined in Office of the Under Secretary of Defense (Comptroller), Department of Defense, *Directive 7045.14, The Planning, Programming, Budgeting, and Execution Process (PPBE)* Process, January 25, 2013 (Incorporating Change 1, August 29, 2017). For more information on the FYDP see CRS In Focus IF10831, *Defense Primer: Future Years Defense Program (FYDP)*, by Brendan W. McGarry and Heidi M. Peters For more information on the PPBE process, see CRS In Focus IF10429, *Defense Primer: Planning, Programming, Budgeting, and Execution (PPBE)*

priorities two, three, and four years into the future to ensure that upcoming projects have time to develop and meet the planning and design requirements (and accurate cost estimates) when they are ready for inclusion in an authorization and appropriation request.

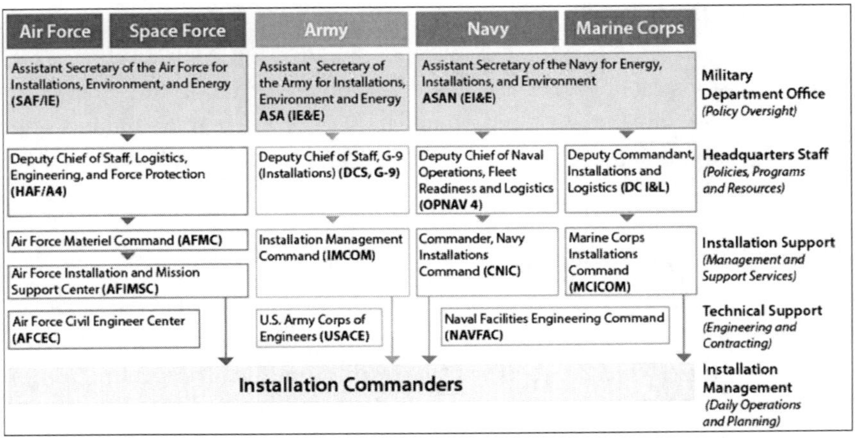


Source: CRS graphic based on DOD component documentation.

Notes: The technical support agencies (USACE, NAVFAC and AFIMSC) provide technical expertise in developing MILCON projects, but they are not directly involved in the planning and prioritization of MILCON budgets.

**Figure 3.** Military Construction Oversight Structures.

## Department of the Army

The Army's single agency responsible for worldwide installation management is Installation Management Command (IMCOM), which is a major subordinate command of Army Materiel Command (AMC).[71] In 2016, the Department of the Army reorganized IMCOM's two domestic regions—known then as the Atlantic Region and Central Region—into its current structure of three functionally aligned directorates that are co-located with three major Army commands.

---

*Process*, by Brendan W. McGarry and CRS Report R47178, *DOD Planning, Programming, Budgeting, and Execution (PPBE): Overview and Selected Issues for Congress*, by Brendan W. McGarry.

71 Army Material Command, AMC Major Subordinate Commands, at https://www.amc.army.mil/Organization/Major- Subordinate-Commands/. Also see Susan A. Merkner, "IMCOM transitions to Army Materiel Command," *The Real McCoy*, March 22, 2019, at https://static.dvidshub.net/media/pubs/pdf_46766.pdf.

Those directorates include IMCOM Readiness, which is co-located with Army Forces Command (FORSCOM); IMCOM Training, co-located with Army Training and Doctrine Command (TRADOC); and IMCOM Sustainment, co-located with AMC.[72] The two overseas directorates—IMCOM Europe and IMCOM Pacific—remain regionally aligned and are located in Sembach Kaserne, Germany, and Fort Shafter, Hawaii, respectively.

The IMCOM directorates work with local installation commanders to develop MILCON requests, which the directorate compiles, prioritizes, and forwards to IMCOM's headquarters, located at Fort Sam Houston in San Antonio, Texas. The headquarters consolidates and prioritizes the project lists from across the Army and submits a list for further review by the Deputy Chief of Staff (DCS), G9 Installations, the service's senior officer responsible for setting installations- related policy and administering installation resource programming.[73]

The U.S. Army Corps of Engineers (USACE) supports the process of developing the DD Form 1391 and offering architectural, engineering, and construction services and support in the development and use of new facilities.[74] The Centers of Standardization (COS) is the USACE element assigned to assist Army headquarters. The COS advises senior Army leaders on the technical aspects of the MILCON program.[75]

As part of DOD's annual budgeting process, the Army's G-9 office forwards a final list of MILCON requests to the Army's Deputy Chief of Staff, G-8 (Financial Management), which assesses and integrates the installations-related resourcing requirements into the overall budget of the Army.

At this point, the Army's construction requests shift to the civilian Army secretariat (the staff supporting the Secretary of the Army). Responsibility for further review and consideration falls to the Assistant Secretary of the Army for Installations, Energy, and Environment (ASA(IE&E)).

Once approved, the Assistant Secretary forwards the Army's proposed construction projects to OSD, where senior officials consolidate the Army's

---

[72] U.S. Army, "IMCOM transforms regions into new directorates," press release, January 24, 2017, at https://www.army.mil/article/181181/imcom_transforms_regions_into_new_directorates.

[73] U.S. Army, Deputy Chief of Staff, G-9 (Installations), About Us website, at https://www.acsim.army.mil/about.html.

[74] U.S. Army, U.S. Army Corps of Engineers, "Fact Sheet," at https://www.usace.army.mil/Media/Fact-Sheets/Fact-Sheets-View/Article/475469/installation-support/.

[75] U.S. Army, Department of the Army Pamphlet 420–1–2, "Army Military Construction and Nonappropriated-Funded Construction Program, Development and Execution," at https://armypubs.army.mil/epubs/DR_pubs/DR_a/pdf/web/PAM%20420-1-2.pdf.

list of proposed projects with those of the other services and defense agencies into a prioritized DOD master list.

## Department of the Air Force[76]

Air Force installation commanders and Major Commands (MAJCOMs) provide MILCON requests to the Air Force Installation and Mission Support Center (AFIMSC), a subordinate unit of the Air Force Materiel Command (AFMC). The AFIMSC helps develop the justifications for individual projects, compile the DD Form 1391s, and create an Integrated Priorities List (IPL) for the Air Force.[77] These tasks involve working through the Air Force Corporate Structure (AFCS) and with the Facility Management Division of the Air Staff (HAF/A4).[78] HAF/A4 Civil Engineer, Facilities Division reviews MILCON programs, integrates Total Force (Active Air Force, Air National Guard, Air Force Reserve) MILCON projects into a single program and presents the MILCON program to the AFCS for approval.[79] A list of MILCON projects then moves through an approval process involving the Assistant Secretary of the Air Force for Energy, Installations, and Environment (SAF/IE), and finally to the Assistant Secretary of the Air Force for Budget (SAF/FMB), who is responsible for sending a final list of Air Force MILCON projects to OSD.

The AFIMSC houses the Air Force Civil Engineer Center (AFCEC), which provides engineering services to Air Force installations. AFCEC supports installation-level planning, documents design and construction

---

[76] For more detailed information on how the Air Force manages its MILCON program, see U.S. Air Force, HQ/ AFIMSC/AFCEC, *Program Management Plan for Air Force MILCON Execution*, September 25, 2017, archived at https://www.wbdg.org/FFC/AF/POLICY/PgMP_MILCON_HAF_AFCEC_AFIMSC_Final.pdf.

[77] U.S. Air Force, HQ/ AFIMSC/AFCEC, *Program Management Plan for Air Force MILCON Execution*, September 25, 2017, pp. 4-5, archived at https://www.wbdg.org/FFC/AF/POLICY/PgMP_MILCON_HAF_AFCEC_AFIMSC_Final.pdf.

[78] The Air Force Corporate Structure is a formal body consisting of civilian and military personnel from the Air Staff or Secretariat, Core Function Lead representatives and the major commands (MAJCOMs). It is established under the Air Force Council, which is chaired by the Vice Chief of Staff of the Air Force and the Under Secretary of the Air Force. See U.S. Air Force, *Planning, Programming, Budgeting, and Execution System Training Program Reference Manual*, pp. 43-46, at http://afacpo.com/AQDocs/PPBE.pdf.

[79] Additional information on Air Force MILCON program and project roles and responsibilities can be found in U.S. Air Force, Instruction (AFI) 32-1021, *Planning and Programming Military Construction (MILCON) Projects*, at https://www.wbdg.org/FFC/AF/AFI/afi_32_1021.pdf.

schedules, and oversight of construction contracts.[80] Though AFCEC is not technically a DOD-designated *construction agent* (i.e., an approved DOD activity assigned the design or construction execution responsibilities associated with a service's MILCON program[81]), AFCEC serves as the design/construction manager for active component projects.[82] In many instances, USACE or Naval Facilities Engineering Systems Command (NAVFAC), as the DOD's designated Construction Agents, provide support for Air Force construction projects.

***Space Force***

The U.S. Space Force (USSF) has its own governance process and conducts its own MILCON Working Group, with support from AFIMSC, to develop a USSF IPL that is separate and distinct from the U.S. Air Force IPL. Starting in FY2023, the Department of the Air Force (DAF) allocates to USSF a portion of the requested DAF MILCON budget.[83]

## Department of the Navy

The Navy's installation commanders and their chain of command within Commander, Navy Installations Command (CNIC), initially prioritize Navy MILCON projects. The CNIC is responsible for worldwide management of the Navy's shore installations. Within CNIC, two divisions—-Facilities and Environmental (N4) and Strategy and Future Shore Integrated Requirements, (N5) —coordinate efforts to plan and prepare MILCON priority lists. The Shore Mission Integration Group (SMIG) evaluates regional priority lists to

---

[80] Ibid., p. 5.

[81] DOD, Department of Defense Directive (DODD) 4270.5, *Military Construction*, updated August 31, 2018, at https://www.esd.whs.mil/Portals/54/Documents/DD/issuances/dodd/427005p.pdf.

[82] The MILDEP's *Construction Agents*, according to DOD Directive 4270.5, are the Naval Facilities Engineering Command (NAVFAC) for the Navy and Marine Corps, and the Army Corps of Engineers (USACE) for the Army. Historically, either NAVFAC or USACE have executed design and construction for the Air Force MILCON program, but more in some instances, AFCEC has provided management and oversight via *Design Manager/Construction Manager* (DM/CMs). See U.S. Air Force, Air Force Instruction (AFI) 32-1023, *Designing and Constructing Military Construction Projects*, December 23, 2020, at https://static.e-publishing.af.mil/production/1/af_a4/publication/afi32- 1023/afi32-1023.pdf.

[83] Information provided to CRS by the U.S. Air Force legislative liaison office via email on May 7, 2024, on file with author.

determine the Navy priorities for MILCON. CNIC consolidates, reviews, and prioritizes the projects for submission to the Navy's senior leadership.[84]

The execution agent for Navy MILCON projects is the Naval Facilities Engineering Systems Command (NAVFAC), which operates worldwide to plan, build, and maintain Navy facilities, and deliver base services. NAVFAC is organized under two regional commands—NAVFAC Atlantic in Norfolk, Virginia, and NAVFAC Pacific in Pearl Harbor, Hawaii. NAVFAC reports overseeing a total of 13 component commands.[85] At most Navy bases, NAVFAC has a Resident Officer in Charge of Construction (ROICC) to oversee MILCON projects for the base. NAVFAC helps prepare and refine the DD Form 1391s and provides input on whether individual projects have sufficient planning and design work completed to execute the construction project during the current budget cycles and their authorization and appropriations window. NAVFAC reviews and confirms the readiness and executability of submitted construction projects; NAVFAC does not prioritize projects. Project prioritization is the role of CNIC and the Navy leadership.

NAVFAC's input on the status of a project's planning and design development can be a factor in the Navy's prioritization decisions for the current year budget submissions.

After CNIC compiles a list of prioritized projects (and NAVFAC evaluates the readiness of those projects) CNIC submits the list to the Chief of Naval Operations (CNO) through the Deputy Chief of Naval Operations (Fleet Readiness and Logistics, OPNAV N4) and, more specifically, the Director, Shore Readiness Division (OPNAV N4I). Upon evaluation and approval, the CNO's staff passes the list to the Assistant Secretary of the Navy for Energy, Installations, and Environment (ASN(EI&E)) for submission to OSD.

### *Marine Corps*

The Marine Corps, as a component of the Department of the Navy, has no installation engineering staff; it relies on NAVFAC and the Navy's ROICCs to plan and administer construction on its installations. Throughout the process, NAVFAC supports the Marine Corps in the same way it supports the Navy, by refining the DD Form 1391 documentation for each project and confirming its readiness and executability.

---

[84] CNIC provides oversight on all shore installation management, exercises authority as the Budget Submitting Office for installation support, and acts as the Navy point of contact for program execution oversight and installation policy. See "Commander, Navy Installations Command website, "History of CNIC," at https://www.cnic.navy.mil/About/History/.

[85] For a list of NAVFAC component commands, see https://www.navfac.navy.mil/About-Us/.

At the same time, the Marines do channel their construction requests through their own chain of command for requirements screening and project prioritization. After Marine Corps installation commanders determine construction requirements, the Marine Corps Installations Command (MCICOM) screens and consolidates the projects and their requisite DD Form 1391.[86] MCICOM reviews and prioritizes the projects and forwards the list to their respective Fleet Marine Force (FMF) headquarters.[87] After the FMF headquarters review, proposed MILCON projects move to the Headquarters, Marine Corps, Installations and Logistics (HQMC/I&L) office. There, Marines finalize a list of MILCON priorities and submit that to the ASN(EI&E) for presentation to OSD.

## National Guard

The National Guard (Army and Air) is a state-based reserve component of the armed forces.[88] Unless the National Guard is federalized (i.e., called to active duty according to statutory provision), each state's National Guard is its own distinct military organization, with the governor as its commander in chief and the adjutant general, or TAG, as its senior military officer.[89] Each National Guard is a joint organization under the TAG. Army National Guard (ARNG) and Air National Guard (ANG) MILCON requirements are handled somewhat differently.

In many instances, federal statute requires the federal government and state government to share the cost of construction for state-based Guard facilities, with the federal government contributing no more than 75 percent of

[86] MCICOM is the single authority for all Marine Corps installations matters. It exercises command and control of regional installation commands, establishes policy, exercises oversight and prioritizes resources in order to optimize installation support to the operating forces, tenant commands, and Marine families. See https://www.mcicom.marines.mil/.

[87] There are two FMFs—Atlantic (FMFLANT) and Pacific (FMFPAC)—commanding the deploying Marine Expeditionary Forces. FMFLANT is headquartered at Naval Station Norfolk, Virginia, while FMFPAC is headquartered at Camp H.M. Smith, Hawaii.

[88] For background information on the Reserve Forces, see CRS In Focus IF10540, *Defense Primer: Reserve Forces*, by Lawrence Kapp, at https://crsreports.congress.gov/product/pdf/IF/IF10540.

[89] With the exception of the District of Colombia, territories (including Guam, U.S. Virgin Islands, and Puerto Rico) also have elected governors. For more information on federalization of the National Guard and other Reserve Forces, see CRS In Focus IF10540, *Defense Primer: Reserve Forces*, by Lawrence Kapp.

the total estimated cost of the project.[90] As a result, state governments are often required to provide matching funds equal to at least 25 percent of the cost of a National Guard construction project. There are numerous exceptions to this rule. The federal government may fund up to 100 percent of a project's cost if the Guard's federal mission or other DOD requirement is primarily driving the construction requirement. States may submit requests for federal funding that exceed 75 percent of an individual project through the Army (for the ARNG) or the Air Force (for the ANG).[91]

### *Army National Guard*[92]

Within a state's joint National Guard headquarters organization (sometimes called a Military Department or Department of Military Affairs), the Army section of the headquarters staff includes a Construction and Facilities Management Officer (CFMO) who administers the state's Military Construction Army National Guard (MCNG) program. CFMOs are the installation engineers of the ARNG and the principal advisors to the TAG regarding all real property, facilities, construction, and environmental management programs.[93] CFMOs conduct master planning and develop a long-range construction plan (LRCP). CFMOs also prepare DD Form 1391

---

[90] The unique authorities for funding construction of National Guard facilities are outlined in by 10 U.S.C. §§18231- 18240. Specifically, the state contribution requirement is outlined in §18236, "Contributions to States; other use permitted by States," which states in section (b): "(b) A contribution made for an armory or readiness center under paragraph ... may not exceed the sum of... a percentage of the cost of construction (exclusive of the cost of architectural, engineering and design services) calculated so that upon completion of construction the total contribution (including the contribution for architectural, engineering and design services) equals 75 percent of the total cost of construction (including the cost of architectural, engineering and design services)." This section of the statute also notes the contribution is required "[e]xcept as otherwise agreed when the contribution is made."

[91] Information Paper provided to CRS by the National Guard Bureau, "Army National Guard Major Military Construction (MILCON) – State Share Requirements," dated April 5, 2022. On file with the author.

[92] For more detailed information on Army National Guard MILCON program development and execution, see National Guard Regulation 415-5, *Construction, Army National Guard Military Construction Program Development and Execution*, March 6, 2015, at https://www.ngbpmc.ng.mil/Portals/27/Publications/ngr/ngr%20415-5.pdf?ver=2018-09-07-082540-720.

[93] Sometimes the CFMO is represented by an office within a state military department. In these cases the term CFMO may be understood as "Construction and Facilities Management Office" rather than officer. However, per National Guard Regulation 415-5, responsibility for ARNG MILCON lies with the state-designated officers. See, for example, the Texas Military Department CFMO at https://tmd.texas.gov/construction-and-facilities-management-office.

for each proposed MILCON project.[94] Following the CFMO's review of the projects and submission to the TAG for approval, the TAG sends a final project list (which includes the DD Form 1391s) to the National Guard Bureau (NGB), an agency that acts as the liaison between the states' National Guard headquarters organizations and the Departments of the Army and Air Force.

Within NGB, ARNG staff (such as those in the Installations Division, or ARNG-ILI) review the submitted MILCON projects from all jurisdictions and prioritize those into an Infrastructure Requirements Plan (IRP). The IRP serves as the prioritized list of Army Guard MILCON projects for federal funding. The director of the ARNG provides the IRP to the Army Staff, for review by the DCS, G-9.[95] From this point, the ARNG's MILCON projects list moves through the review and approval process within the Department of the Army and Army Secretary's office. ARNG projects follow the Army scoring process for prioritization within the ARNG portion of the MILCON Total Obligation Authority (TOA) for inclusion the FYDP.[96]

### *Air National Guard*[97]

MILCON projects for the ANG are primarily intended to meet current and new missions stemming from the Department of the Air Force's corporate Air Force strategic basing process. The Base Civil Engineers (BCEs) for each state's ANG installations initiate MILCON projects through a process similar to that of the ARNG. BCEs produce DD Form 1391s for each MILCON project. The Air Staff within the state's joint National Guard headquarters compiles and prioritizes the projects within the state according to the TAG's general guidance. After approving a final list, the TAG sends the state's highest priority ANG projects to the NGB for evaluation by the ANG Headquarters Air Staff.[98] The ANG Air Staff combines and refines the list for

---

[94] National Guard Regulation 415-5, *Construction, Army National Guard Military Construction Program Development and Execution*, March 6, 2015, p. 2, at https://www.ngbpmc.ng.mil/Portals/27/Publications/ngr/ngr%20415-5.pdf?ver=2018-09-07-082540-720.

[95] National Guard Pamphlet 210-20, "Real Property Development Planning Procedures for The Army National Guard," at https://www.ngbpmc.ng.mil/Portals/27/Publications/NGPAM/ngpam%20210- 20.pdf?ver=oRUMEnfm1FpHJToG0Kuskg%3d%3d.

[96] Ibid.

[97] For more detailed information on Air National Guard MILCON program policies, responsibilities, and execution, see Air National Guard Instruction 32-1023, *Criteria and Standards for Air National Guard Design and Construction*, January 21, 2015, at https://www.wbdg.org/FFC/ANG/ANGI/ANGI_32-1023.pdf.

[98] Air Force Policy Directive 32-10, 20 JULY 2020, Civil Engineering Installations and Facilities, see page 4, section 3.4, at https://static.e-publishing.af.mil/production/1/saf_ie/publication/afpd32-10/afpd32-10.pdf#page=4.

the director of the ANG, who approves and forwards a final list of ANG MILCON projects to Headquarters, Air Force (HAF) staff.[99] From that point, the ANG MILCON projects list is integrated into the broader review and approval process for the entire Department of the Air Force. The Facility Management Division, AF/A4CF, reviews MILCON programs, and integrates Total Force (Active Air Force, Air National Guard, Air Force Reserve) MILCON projects into a single prioritized list.[100] The Deputy Assistant Secretary of the Air Force for Installations (SAF/IEI) oversees all Air Force and ANG MILCON programs for the Department of the Air Force.[101]

## Office of Management and Budget (OMB) Review

The joint review of the DOD's annual budget by the OSD and the White House's Office of Management and Budget (OMB) generally occurs between September and December, and focuses on the administration's priorities; aspects of this process may vary year to year depending on current events or leadership preferences.[102] This review includes MILCON programs and the review of DD Form 1391 submissions for projects proposed by OSD.

The highest priority projects that fit within the budgetary guidance provided by the OMB will be included in the President's budget request. Some of the approved projects assigned a lower priority are slated for funding in future years and are typically included in the Future Years Defense Program (FYDP).[103] One reason that some projects may be scheduled in the FYDP

---

[99] Air Force Instruction 32-102, Planning and Programming Military Construction (MILCON) Projects, February 2016, see page 5, at https://www.wbdg.org/FFC/AF/AFI/afi_32_1021.pdf#page=5.

[100] Ibid.

[101] For more detailed information on Air National Guard MILCON program policies, responsibilities, and execution, see Air National Guard Instruction 32-1023, Criteria and Standards for Air National Guard Design and Construction, January 21, 2015, at https://www.wbdg.org/FFC/ANG/ANGI/ANGI_32-1023.pdf#page=7.

[102] CRS In Focus IF10429, Defense Primer: Planning, Programming, Budgeting, and Execution (PPBE) Process, by Brendan W. McGarry.

[103] 10 U.S.C. §221 requires that the Secretary of Defense submit, at about the same time as the President's annual budget request, a "future-years defense program … reflecting the estimated expenditures and proposed appropriations included in that budget. Any such future-years defense program shall cover the fiscal year with respect to which the budget is submitted and at least the four succeeding fiscal years." DOD develops its FYDP according to policies and procedures outlined in Office of the Under Secretary of Defense (Comptroller), Department of Defense, *Directive 7045.14, The Planning, Programming, Budgeting, and Execution Process (PPBE)* Process, January 25, 2013 (Incorporating Change 1, August 29, 2017). For more information on the PPBE process, see CRS In Focus

rather than the current-year budget is their planning and development is not sufficiently developed, raising questions about the reliability of the existing cost estimate or making it unlikely the projects could be executed on the timeline of current appropriations.[104] Therefore, some MILCON projects may survive the screening process and reach high-level review only to be placed in the budgeting queue for a future year's submission. While the FYDP is an indicator of future priorities, it is also possible that a MILCON project's priority may diminish in a later year because budget levels fluctuate, missions change, and the entire construction list is reviewed and reevaluated every year.

## Authorizations, Appropriations, and Enactment[105]

The annual budgeting cycle formally begins with the release of the President's budget request (traditionally in February, but sometimes later), which includes a list of military construction projects prioritized by the DOD and White House. The annual process normally ends with enactment of two key pieces of legislation – the NDAA and the MILCON-VA Act. See Figure 5.

A list of the military construction projects authorized each year along with the specific amount of funding authorized is contained in the NDAA. A similar but separate list of military construction appropriations (i.e., projects that are to receive budgetary authority) is contained in the explanatory statement that accompanies the annual MILCON-VA Act.[106]

The two pieces of legislation may also provide policy guidance relevant to MILCON matters. In addition to what is included in the bills themselves, committee reports accompanying the armed services and appropriations bills may provide useful information. For example, the accompanying reports may direct military leaders to provide reports to Congress about certain matters;

IF10429, *Defense Primer: Planning, Programming, Budgeting, and Execution (PPBE) Process*, by Brendan W. McGarry and CRS Report R47178, *DOD Planning, Programming, Budgeting, and Execution (PPBE): Overview and Selected Issues for Congress*, by Brendan W. McGarry.

[104] CRS In Focus IF10831, *Defense Primer: Future Years Defense Program (FYDP)*, by Brendan W. McGarry and Heidi M. Peters.

[105] For more information on the MILCON appropriation and obligation process, see GAO, *Defense Infrastructure: Action Needed to Increase the Reliability of Construction Cost Estimates*, GAO-18-101, March 2018, at https://www.gao.gov/assets/700/690892.pdf.

[106] For general information on the federal budgeting, appropriations, and authorization processes, see CRS Report 98- 721, *Introduction to the Federal Budget Process*, coordinated by James V. Saturno, and CRS Report RS20371, *Overview of the Authorization-Appropriations Process*, by Bill Heniff Jr.

may set guidance for funding policies or express the views of the defense committees on policy matters not directly addressed in the legislation.

While the committees and subcommittees exercise jurisdiction over the legislation, they also typically provide opportunities for individual Members of Congress not assigned to the committees of jurisdiction to provide input regarding specific MILCON projects. During the period when the committees are drafting the legislation, the committees may issue guidance for Members wishing to submit input for on MILCON matters. The committees' guidance might suggest additional documentation that could prove useful to the committee in its deliberations, or may also set a deadline for submission of such information. Often that deadline is set for late March, when the subcommittee staffs begin to prepare for markup of the necessary legislation.

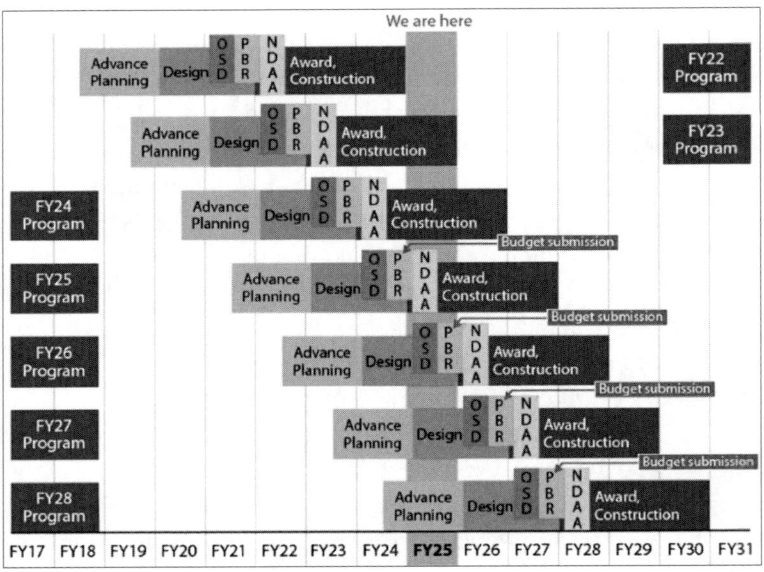


Source: Derived from Department of the Army briefing information provided to CRS. On file with the author.

Notes: The blue-green box labeled "OSD" illustrates the stage at which the Office of the Secretary of Defense (OSD) reviews and provides input on the annual MILCON budget requests. The yellow box labeled "PBR" reflects the process of incorporating Defense Department requests into the final President's budget request (PBR) that the White House submits to Congress each year.

**Figure 4.** Military Construction Planning, Programming, Budgeting and Execution Cycles.

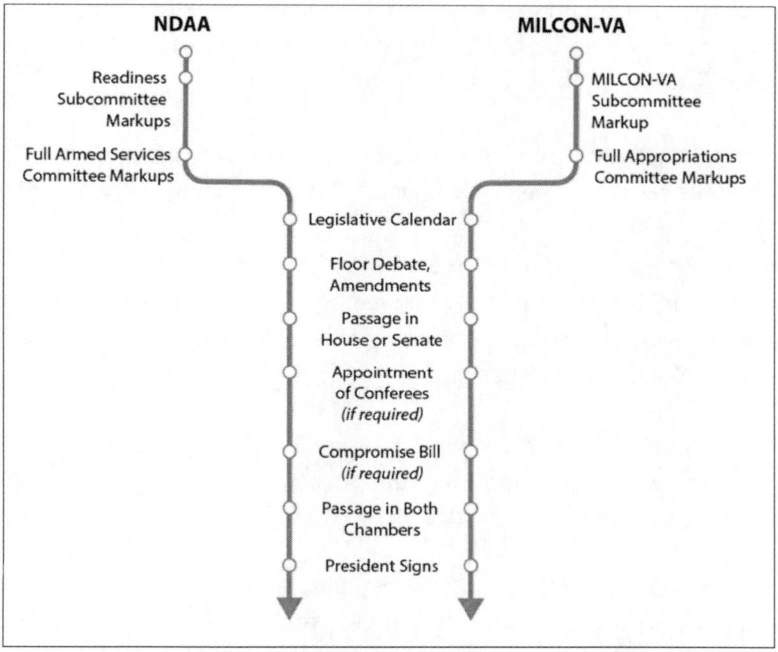


Source: CRS graphic. Information drawn from House and Senate committee websites, CRS Report RS21363, Legislative Procedure and Process Resources for Congressional Staff, by Jennifer E. Manning and Maura Mullins.

**Figure 5.** Legislative Process/Congressional Operations.

## Community Project Funding and Congressionally Directed Spending Items

In some instances, Congress may insert into the authorization and appropriations acts MILCON projects that were not included in the President's budget request. This often results from a specific request from a Member, known as "earmarks" — formally referred to as "Congressionally Directed Spending" in the Senate and "Community Project Funding Items" in the House of Representatives. In consideration of specific military construction projects that the DOD and White House did not prioritize for the current budget cycle, Congress often defers to a set of guidelines known as the "McCain Rules." These rules are not codified in statute. They were approved by the Senate in a nonbinding resolution included in the FY1995 NDAA (P.L. 103-337; §2856). The "McCain Rules" state that consideration of authorization and appropriations for any military construction project not included in the annual budget submission from the White House should be limited to MILCON projects that are:

- considered essential to the DOD's national security mission;
- consistent with past actions under the Base Realignment and Closure Act (BRAC);
- in the services' Future Years Defense Program;
- executable in the year they are authorized and appropriated; and
- offset by reductions in other defense accounts, through advice from the Secretary of Defense.

DOD applies these rules in the formulation of the annual report to Congress on unfunded priorities required by 10 U.S.C. §222a, generally known as the unfunded priorities lists (UPL) or unfunded requirements (UFRs).[107]

The defense committees may issue guidance to Members as a part of the annual budget cycle.[108] For example, in March 2022, the House Appropriations Committee's subcommittee on Military Construction, Veterans Affairs and Related Agencies issued a "Fiscal Year 2023 Member Request Guide." That guidance stated in part: "The eligible lists of community projects are those that are submitted to Congress by the Secretary of Defense or his/her designee. Projects that only appear on a project list provided by a base commander, as opposed to the Secretary or his/her designee, will not be accepted."[109] In addition, the guidance describes such eligible lists as the Future Years Defense Program (FYDP) or Unfunded Requirements/Unfunded Priorities Lists (UFRs/UPLs).

---

[107] For additional information, see CRS In Focus IF11964, *Defense Primer: Defense and Intelligence Unfunded Priorities*, by Brendan W. McGarry.

[108] For example, the Senate Committee on Appropriations issued guidance for the FY2025 budget cycle. https://www.appropriations.senate.gov/imo/media/doc/fy2025_appropriations_requests_general_guidance1.pdf. Also, the House Committee on Appropriations, Subcommittee for Military Construction, Veterans Affairs and Related Agencies issued guidance specific to MILCON-related requests. https://appropriations.house.gov/sites/evo-subsites/republicans-appropriations.house.gov/files/evo-media-document/fy25-military-construction-veterans-affairs- and-related-agencies-guidance.pdf.

[109] Fiscal Year 2023 Member Request Guide Subcommittee on Military Construction, Veterans Affairs and Related Agencies, at https://appropriations.house.gov/sites/democrats.appropriations.house.gov/files/FY23%20MilConVa%20Member%20Request%20 Guide.pdf.

The practice of Congress inserting earmarks was suspended in 2011 but resumed in 2021. Members typically have to meet certain requirements under Senate and House rules to have their requests included as provisions in defense authorization and appropriations acts.[110] Such requirements included Members posting requests online and certifying that they had no financial interest in the projects.[111]

In total, 142 additional MILCON projects totaling nearly $1.56 billion were included in the MILCON-VA Appropriations Act for 2024, (Division A of P.L. 118-42).[112] Before enactment of a MILCON-VA appropriations act that includes projects not identified in the budget request, DOD may identify its priority military construction projects that had not yet been funded by Congress, had undergone some preliminary design, and would be ready for execution within the fiscal year.[113] After members submit requests, Congress reviews the potential projects, discusses them with DOD officials, and decides which ones to include in the annual appropriations act.[114]

## MILCON Execution

Execution of the construction for MILCON projects may begin after the President signs the authorization and appropriations legislation into law. Construction encompasses several phases, including the solicitation of bids from prospective general contractors, awarding of contracts, construction of the facility or infrastructure, outfitting of equipment, and any warranty period provided by contractors involved in the project.

---

[110] See, for example, guidance promulgated by the House Appropriations Committee, Transparency website, "Appropriations Requests," at https://appropriations.house.gov/transparency/appropriations-requests-2023. For more background, see CRS Report RS22866, *Earmark Disclosure Rules in the House: Member and Committee Requirements*, by Megan S. Lynch and CRS Report RS22867, *Earmark Disclosure Rules in the Senate: Member and Committee Requirements*, by Megan S. Lynch.

[111] GAO, *Tracking the Funds: Specific Fiscal Year 2022 Provisions for Department of Defense*, GAO-23-105914, October 4, 2022, at https://www.gao.gov/products/gao-23-105914.

[112] See the table, "Community Project Funding/Congressionally Directed Spending Items," in the joint explanatory statement to accompany the Military Construction, Veterans Affairs, and Related Agencies Appropriations Act, 2024, (Division A of P.L. 118-42) published in the Congressional Record for March 5, 2024, p. S1235, at https://www.congress.gov/118/crec/2024/03/05/170/39/CREC-2024-03-05.pdf#page=184.

[113] For example, see GAO, Tracking the Funds: Specific Fiscal Year 2022 Provisions for Department of Defense, GAO-23-105914, October 4, 2022, at https://www.gao.gov/products/gao-23-105914.

[114] GAO, *Tracking the Funds: Specific Fiscal Year 2022 Provisions for Department of Defense*, GAO-23-105914, October 4, 2022, at https://www.gao.gov/products/gao-23-105914.

The responsibility for executing and managing MILCON projects typically falls to one of the two DOD Construction Agents (DCA), NAVFAC or USACE, or to a designated Design Manager/Construction Manager (DM/CM).[115] However, ARNG and ANG headquarters bureaus manage their respective MILCON projects separately.[116] If additional architectural and engineering work is needed for the project, the DCA may complete the final design in-house, or the DCA may choose to contract with an external design firm to finalize the planning and design work. In instances where the service-designated construction agents/managers are not able to perform the design or construction work themselves, they are responsible for preparing, advertising, evaluating, and awarding design and construction contracts. Both NAVFAC and USACE have in-house design capabilities. Their MILCON workloads generally exceed their capacity, thus requiring architect/engineer (A/E) contracts for preparing some construction contract documents. With limited exceptions, notably U.S. Naval Construction Battalions (or CBs, known as "Seabees"), the armed services do not have in-house construction capabilities.

Therefore, NAVFAC and USACE typically hire one or more general contractors to execute construction. These contracts are awarded through a specific Request for Proposal (RFP) process.[117]

Typically, the DCAs opt for one of two general contracting approaches, Design-Bid-Build (DBB) or Design-Build (DB), to execute a MILCON project. The difference between these two methods will affect the type of RFP process used. For DBB projects, the full design is prepared either by in-house resources or via an A/E contract. This results in a full set of drawings and

---

[115] Traditionally, the USACE has supported the Army and NAVFAC supports the Navy and Marine Corps. Historically, either NAVFAC or USACE have executed design and construction for the Air Force MILCON program. In some instances, AFCEC has taken on these roles and responsibilities and is designated by the Air Force as its Design Manager/Construction Manager (DM/CM). See U.S. Air Force, Air Force Instruction (AFI) 32-1023, Designing and Constructing Military Construction Projects, December 23, 2020, at https://static.e- publishing.af.mil/production/1/af_a4/publication/afi32-1023/afi32-1023.pdf. Also, while this division of responsibility usually holds for construction within the United States, overseas construction responsibility is typically governed by the project's location. DOD Directive 4270.5 designates specific construction agents by geographical area. Unspecified minor military construction may be executed by the Secretary concerned or, in the case of a Defense Agency or a DOD Field Activity, by the secretary concerned of the MILDEP having jurisdiction over the real property facility.

[116] The Army and Air National Guard are authorized to use their United States Property and Fiscal Officer (USPFO) to perform the actions of the DCA.

[117] For more information on the RFP process, see Step 6: Request for Proposal Preparation and Evaluation in the report: RAND Corporation, Obtaining Life-Cycle Cost-Effective Facilities in the Department of Defense, 2013, p. 17, at https://www.rand.org/pubs/research_reports/RR169.html.

specifications on which the construction contractors may submit bids. This design strategy is most common when the government may want to provide significant input into the design process. For DB projects, the contractor is responsible for hiring and managing the design firm to complete the design. This type of contract is more typical for lower complexity projects. The executing agency will decide which contracting method is in the best interest of the government and is the most feasible means to get a project completed in a timely manner within budget.[118]

## Contract Award, Contract Changes, and Construction Completion

After a project is authorized and funding is appropriated, the DCA may begin soliciting bids through the RFP process in accordance with the project's DD Form 1391. The DCA will continue to manage the project and monitor construction throughout the process.

After construction contracts are awarded, construction contracts frequently need changes. For MILCON projects, in particular, extended timelines and unexpected changes to military missions can amplify the need for changes to original contracts. These changes can be due to oversights in the design process, updated regulations, unforeseen site conditions or mission changes that result in revised military requirements. The DCA must evaluate each potential change to determine if it is within the scope of the project and contract prior to the contracting officer formalizing a modification to the contract. These changes often lead to increased costs. For this reason, DOD adds a standard contingency rate of 5% to initial cost estimates and programs it into the original budget request.[119] If that contingency funding is insufficient to cover the project's needs, there are mechanisms for increasing the project's funding. Mechanisms include increasing the authorization through a reprograming request in accordance with the DOD Financial Management Regulation or a formal request for additional funding authorization through

---

[118] The federal regulations concerning Design-Bid-Build projects are outlined in portions of Title 40 of the *United States Code* (§1101 et seq.), and those relevant to Design-Build projects are in 10 U.S.C. §2305a. Both statutes apply to federal building contracts in general and are not unique to the MILCON process.

[119] Unified Facilities Criteria (UFC) Programming Cost Estimates for Military Construction, see page 8, section 4-6.1, at https://www.wbdg.org/FFC/DOD/UFC/ufc_3_730_01_2011_c2.pdf.

the next annual NDAA and MILCON-VA appropriations bill.[120] The mechanisms for increasing the authorization are outlined in 10 U.S.C. §2853 and are based on specified thresholds.

## Contract Changes: Cost Increase Thresholds

In some instances, the actual cost of executing a military construction project turns out to be less than the amount of money originally authorized and appropriated for that project. This leftover funding is sometimes referred to as "bid savings." In other instances, a project may ultimately cost more than the amount originally authorized and appropriated by Congress.

Under 10 U.S.C. §2853, the Secretary concerned must notify the appropriate congressional committees if a project has cost increases or decreases of more than 25% of the total authorized cost of the project or 200 percent of the minor construction project ceiling, whichever is lower. (This is often referred to as a "2853 notification.") The notification must include: the amount of the cost increase; the reasons for the increase; description of the funds to be used to cover the cost increase, and certification from the Secretary concerned that the cost increase is sufficient to meet the mission requirement identified in the justification data provided to Congress as part of the request for authorization of the project.

In addition to or distinct from statutory notification requirements in 10 U.S.C. §2853, DOD may seek to have bid savings or other unobligated money transferred, or "reprogrammed" from one project to another. The requirements for reprograming money depend on several factors. Each year, generally, the report accompanying the annual MILCON-VA appropriation bill includes language defining the threshold requiring prior approval for reprogramming funds within a military construction appropriations account.[121] In FY2024, the threshold was set at $6 million or 25 percent of the original funded amount, whichever is less.[122] This threshold helps determine the process for seeking additional money: Key elements of that process include the following:

---

[120] Under 10 U.S.C. §2853, the cost authorized for a military construction project may be increased or decreased by not more than 25% of the total authorized cost of the project or 200% of the minor construction project ceiling specified in section 2805(a), whichever is less. The current project ceiling specified in section 2805(a) is $9 million.

[121] Ibid.

[122] The threshold was set for FY2024 in the joint explanatory statement to accompany the Military Construction, Veterans Affairs, and Related Agencies Appropriations Act, 2024, (Division AJ of P.L. 118-42) published in the Congressional Record for March 5, 2024, p.

- For project cost increases within this threshold, Below Threshold Reprogramming (BTR) is permitted.[123] BTRs require internal service processing only, though they must still be funded with bid savings within the applicable MILCON appropriations.[124]
- If the required project cost increase exceeds the BTR limit and bid savings are available to cover the cost overrun, Congress requires a formal reprogramming of funds to assign available bid savings to the project. These Prior Approval (PA) reprogramming actions are typically routed from the service, through the Office of the Under Secretary of Defense Comptroller (OUSD(C)) for approval, and then sent to the appropriate congressional committees for final approval.[125]
- If the required cost increase exceeds the BTR limit and bid savings are not available, the DOD may submit a request for the additional money to complete the project in the next budget cycle. This can be included as a specific line item in the next annual NDAA and MILCON-VA appropriations bill as a "Cost to Complete (CTC), assuming that the funds can be allocated within the military department's total budget authority.

## Contract Changes: Scope Change Thresholds

If the Secretary concerned determines that the scope of a defined facility within a project must be increased or decreased for some reason, the change may require, depending on the extent of the increase or decrease, the Secretary concerned to notify the appropriate congressional committees or seek additional authorization.[126]

If the increase in scope is 10% or less of the original scope reported in the DD Form 1391, the Secretary concerned may approve that increase with a

---

S1224, at https://www.congress.gov/118/crec/2024/03/05/170/39/CREC-2024-03-05.pdf#page=173.

[123] See DOD, *Financial Management Regulation* (7000.14-R), Volume 3, Chapter 6, "Reprogramming of DoD Appropriated Funds," at https://comptroller.defense.gov/Portals/45/documents/fmr/current/03/03_06.pdf.

[124] "Bid savings" refers to any money remaining in the MILCON budget that results from contracts that ultimately (after implementation of the federal contracting process) cost less than the amount that Congress authorization and appropriated to fund the project.

[125] See DOD, *Financial Management Regulation* (7000.14-R), Volume 3, Chapter 7, "Reprogramming of Military Construction and Family Housing Appropriated Funds," at https://comptroller.defense.gov/Portals/45/documents/fmr/current/03/03_07.pdf.

[126] See 10 U.S.C. §2853, "Authorized cost and scope of work variations."

required notification of the appropriate congressional committees and a 14-day waiting period. If the scope increase required exceeds 10%, OSD must submit a request to Congress for approval in the NDAA during the budget cycle with special authorization language to be included under Division B for the service in question.

Section 2853 also limits potential decreases in the scope of a MILCON project.[127] If the scope of work for a military construction project is reduced by more than 25% from the amount specified for that project, the Secretary concerned must grant a waiver for the reduction and is required to notify the appropriate congressional committees.[128] The notification should describe the reduction in the scope of work and the reasons for the decrease; and certify that the project in its reduced scope can still meet the mission requirement originally identified in the justification data provided to Congress. The Secretary's waiver takes effect after a 14-day waiting period following the notification of the appropriate congressional committees.

## Completion

Upon construction completion, the facility is inspected and turned over to the customer for outfitting of any specialized equipment or furnishings not included in the main construction contract. During the transfer of the facility, the DD Form 1354 documents the components of the construction with quantity and values, and provides the basis for the value of the property in the real property database system for the installation. Following turnover of the facility, the DCA continues to monitor the facility for any warranty issues prior to financial closeout of the project.

[127] 10 U.S.C. § 2853 defines the term "scope of work" to refer to the function, size, or quantity of a facility or item of complete and useable infrastructure contained in the justification data provided to Congress as part of the request for authorization of the project, construction, improvement, or acquisition.

[128] The law limits the military's option to scale back the scope of the project to reduce costs. 10 U.S.C. §2853 states: "Any reduction in scope of work for a military construction project shall not result in a facility or item of infrastructure that is not complete and useable or does not fully meet the mission requirement contained in the justification data provided to Congress as part of the request for authorization of the project, construction, improvement, or acquisition."

# Appendix A. Selected Statutes Governing Military Construction and Repair Authorities

**Table A-1.** Summary of selected military construction and repair authorities

| Authorities | Criteria | Limitations | Congressional Reporting Requirement | Waiting Period | Source of Funding |
|---|---|---|---|---|---|
| Military Construction *(10 U.S.C. §2802)* | Provides funding for military construction projects, land acquisitions, and defense access road projects | Projects costing more than $9 million are subject to specific authorization in law and budgetary authority through annual budget cycle | As part of the annual budget process, the Secretary concerned must provide information about the project's description, justification and cost estimate | No waiting period for projects authorized in law | MILCON appropriations as specifically authorized in law |
| Emergency Construction *(10 U.S.C. §2803)* | Provides funding for projects vital to the national security or to the protection of health, safety, or the quality of the environment; *and* So urgent that deferral of the project for inclusion in the next MILCON authorization act would be "inconsistent with national security or the protection of health, safety, or environmental quality" | Total obligations under this authority may not exceed $50 million for any fiscal year (for each Secretary concerned) | Statute requires Secretary concerned to notify Congress when using this authority and to provide a justification for the project, cost estimate, source of funding and rational for using this authority | 5 days from notification | Unobligated funding from MILCON appropriations |
| Contingency Constructione *(10 U.S.C. §2804)* | Provides funding if the Secretary of Defense determines that deferral of the project for inclusion in the next MILCON authorization act would be | Subject to unique DOD approval process involving Joint Chiefs of Staff | Statute requires the Secretary of Defense to notify Congress when using this authority and to provide a justification, a cost | 7 days from notification | MILCON appropriations as specifically authorized for the "Contingency Construction" account |

**Table A-1.** (Continued)

| Authorities | Criteria | Limitations | Congressional Reporting Requirement | Waiting Period | Source Of Funding |
|---|---|---|---|---|---|
| | "inconsistent with national security or national interest" | | estimate and explanation for using this authority | | |
| Unspecified Minor Construction[a] (10 U.S.C. §2805) | Provides funding for Military Construction projects if total cost does not exceed $9 million | If project costs more than $750,000, statute requires prior approval from the Secretary concerned | If project costs more than $4 million, statute requires Secretary concerned to notify Congress, provide justification and cost estimate | 14 days from notification for projects with cost exceeding notification threshold | MILCON appropriations as specifically authorized for Unspecified Minor Military Construction accounts<br><br>If project costs less than $4 million, statute allows Secretary concerned to use O&M appropriations<br><br>Money from Working Capital Funds may be used for projects revitalizing the Defense Industrial Base[b] |
| Laboratory Revitalization (10 U.S.C. §2805(g)) | Provides funding for projects for the purpose of revitalization and recapitalization of DOD laboratories, which includes facilities conducting "test and evaluation activities" and | The maximum amount of RDT&E funds appropriated or otherwise made available for military construction projects in any fiscal year for military construction projects is $150 million. | Major MILCON projects using RDT&E funding under this statute must requested and justified through the annual defense budget process. | No waiting period when authorized and appropriated through annual defense budget process | RDT&E appropriations |

| Authorities | Criteria | Limitations | Congressional Reporting Requirement | Waiting Period | Source Of Funding |
|---|---|---|---|---|---|
| | "research, engineering and development centers" | | | | |
| Laboratory Revitalization (10 U.S.C. §2805(d)) | Provides funding for minor military construction projects (those costing less than $9 million) for the purpose of revitalization and recapitalization of DOD laboratories, which includes facilities conducting "test and evaluation activities" and "research, engineering and development centers" | If project costs more than $750,000, statute requires prior approval from the Secretary concerned | If project costs more than $4 million, statute requires Secretary concerned to notify Congress, provide justification and cost estimate | 14 days from notification | MILCON appropriations as specifically authorized for Unspecified Minor Military Construction accounts<br><br>O&M appropriations, or amounts appropriated under 10 U.S.C. §4123[g] |
| Architectural and Engineering Services and Construction Design (10 U.S.C. §2807) | Provides funding for architectural and engineering services, as well as construction design performed in connection with MILCON and family housing projects[c] | Funding may not be used for activities related to advance planning[d] | If services for individual projects cost more than $1 million, statute requires Secretary concerned to notify Congress, describing the scope of the project and estimated cost of services | 14 days from notification for projects with cost exceeding notification threshold | MILCON appropriations as specifically authorized for "Planning & Design" accounts |
| Construction Authority in the Event of a Declaration of War or National Emergency (10 U.S.C. §2808) | Provides funding in the event of a declaration of war, or declaration of a national emergency by the President requiring use of the armed forces for projects "necessary to support the use of such armed forces" | Total obligations under this authority may not exceed $500 million for any fiscal year and may not exceed $100 million for projects undertaken within the United States. | Statute requires the Secretary of Defense to notify Congress when using this authority and provide an explanation of "how each construction project directly supports the immediate security, logistical, or short-term housing and ancillary supporting facility needs of the | 5 days from notification | Unobligated funding from MILCON appropriations (excluding funds appropriated for family housing) |

**Table A-1.** (Continued)

| Authorities | Criteria | Limitations | Congressional Reporting Requirement | Waiting Period | Source Of Funding |
|---|---|---|---|---|---|
| | | Authority terminates "at the end of the war or national emergency" | members of the armed forces used in the national emergency."<br><br>Statute also requires notification of cost estimates and information about any projects that were canceled or modified to provide funding for the emergency projects. | | |
| Repair of Facilities (10 U.S.C. §2811) | Provides funding for repairs or restoration of a real property facility or system for its original or designated functional purpose;<br>or<br>Conversion of a real property facility or system to a new functional purpose without increasing its external dimensions | Statute prohibits construction of new facilities or additions to existing facilities<br><br>If project costs more than $7,500,000, statute requires prior approval from the Secretary concerned | If project costs more than $7.5 million, statute requires Secretary concerned to notify Congress regarding the project's Justification and total cost estimate;<br><br>If repair costs exceed 75% of the estimated replacement cost, notification must include an explanation of why the facility's replacement is not in the best interest of the Government | None | O&M appropriations subject to availability |
| Military Construction Projects for Energy Resilience, Energy | Provides funding for projects to improve an installation's energy resilience or energy security, or to improve energy conservation | Obligated funding available from O&M accounts under this authority is limited to | Statute requires Secretary seeking to use O&M funds for this purpose to notify Congress and provide justification for the | 7 days from notification | MILCON appropriations as specifically authorized in law under §2802 or |

| Authorities | Criteria | Limitations | Congressional Reporting Requirement | Waiting Period | Source Of Funding |
|---|---|---|---|---|---|
| Security, and Energy | | $100 million in any fiscal year. | project, cost estimates, the source of the O&M funds to be | | Secretary may use O&M accounts subject to Congressional notification |
| Conservation (10 U.S.C. §2914) | | | used, and, in the case of a military construction project for energy conservation, the most current projected savings-to-investment ratio. | | |
| Military Installation Resilience Projects (10 U.S.C. §2815) | May be carried out on a military installation or outside of a military installation if the Defense Secretary determines that the project would preserve or enhance the resilience of a military installation, military | Must be carried out in accordance with §2802 and Secretary must provide the rationale for how the project would enhance military installation resilience; enhance mission | Statute requires the Secretary of Defense to notify Congress when using this authority and to provide a rational for using this authority<br><br>If using O&M funding, statute requires Defense Secretary to | 14 days from notification | MILCON appropriations as specifically authorized in law under §2802 or Secretary may use O&M accounts subject to Congressional notification |
| | facility or community infrastructure necessary to maintain or rapidly reestablish an installation's mission-essential functions | assurance; support mission critical functions; or address known vulnerabilities or If the Secretary concerned seeks to use O&M funding for a project not previously authorized in law, the Secretary concerned must certify that that waiting for next MILCON authorization act would be | notify Congress and provide cost estimates, source of funding and certification that waiting for next MILCON authorization act would be inconsistent with national security or the protection of health, safety, or environmental quality,<br><br>Statute also requires the Secretary of Defense to submit | | |

**Table A-1.** (Continued)

| Authorities | Criteria | Limitations | Congressional Reporting Requirement | Waiting Period | Source Of Funding |
|---|---|---|---|---|---|
| | | inconsistent with national security or the protection of health, safety, or environmental quality | an annual report on the status of projects[f] | | |
| Authority for certain construction projects in friendly foreign countries (10 U.S.C. §2817) | Provides funding for construction projects in friendly foreign countries at an air, or sea port of debarkation, rail or other logistics support location | Individual projects may not exceed $15 million and total obligations under this authority may not exceed $200 million in any fiscal year | For construction projects with a total cost that exceeds the amounts authorized for unspecified minor military construction projects under section 2805, this statute requires the Secretary of Defense to notify Congress when using this authority along with a supporting rationale | 30 days from notification | O&M appropriations subject to availability |
| Restoration or replacement of damaged or destroyed facilities (10 U.S.C. §2854) | Provides funding for repair, restore, or replace a facility, including a family housing facility, that has been damaged or destroyed.<br>O&M funding may be used if damage or destruction to the facility was the result of a natural disaster or a terrorism incident | Total obligation in O&M funds under this authority may not exceed $100 million in any fiscal year. | Statute requires the Secretary concerned to notify Congress if the cost of the repair project exceeds $9 million. Notification should include cost estimates, the source of funds for the project, and of the justification for carrying out the project under this section | Waiting period varies depending on the type of funding used (MILCON vs. O&M) | Unobligated funding from MILCON appropriations or O&M appropriations subject to availability |

Source: Title 10 of the United States Code, Congressional Research Service.

Notes:

[a] Under §2805(f), the Secretary concerned may adjust the dollar limitations specified in 10 U.S.C. §2805 (Unspecified minor construction) for unspecified minor military construction projects inside the United States to reflect the area construction cost index for military construction projects (not to exceed $14M).

[b] 10 U.S.C §2208 provides temporary authorization for the use of working capital funds to pay for minor military construction projects that aims to "revitalize or recapitalize defense industrial base facilities." The statute defines defense industrial base facilities as "any Defense Department depot, arsenal, shipyard or plant located within the United States. That provision is scheduled to expire at the end of FY2025, per Sect. 372 of P.L. 117-263.

[c] Includes projects undertaken in connection with the authority provided under 10 U.S.C. § 2854 (Restoration or replacement of damaged or destroyed facilities).

[d] Funding under this statute provides for project-specific architectural and design work. Activities related to the evaluation of the initial military requirement, known as advance planning, are funded through O&M accounts. For more information about advance planning, see pages 6 and 14 of this report.

[e] Contingency Construction funding is not always included in the annual budget. For example, the conference report (H.Rept. 112-331) accompanying the Consolidated Appropriations Act, 2012 (P.L. 112-74) listed $10 million for contingency construction, within the Military Construction, Defense-Wide account. See H.Rept. 112-331, p. 1283. Since 2012, the defense legislation has not included appropriations for contingency construction under 10 U.S.C. §2804.

[f] Statute requires the Secretary of Defense to submit an annual report to the congressional defense committees not later than 90 days after the end of each fiscal year until December 31, 2025 on the status of the planned and active projects carried out under this section (including completed projects). Reports should include the title, location, a brief description of the scope of work, the original project cost estimate, and the current working cost estimate, and explanation for how the project will or has enhanced military installation resilience.

[g] 10 U.S.C. §4123 provides funding for use at the discretion of the directors of defense laboratories to fund innovation and development programs, to potentially include construction of infrastructure, in support of the defense laboratories' military missions.

# Appendix B. Example DD Form 1391: Joint Base Lewis-McChord Barracks

Included in the figures below is the DD Form 1391 for a $49 million barracks project for Joint Base Lewis-McChord, which the Army submitted to Congress as a supporting document in its FY2023 MILCON budget request.

| 1. COMPONENT | | 2. DATE |
|---|---|---|
| Army | FY 2023 MILITARY CONSTRUCTION PROJECT DATA | 01 APR 2022 |

| 3. INSTALLATION AND LOCATION | 4. PROJECT TITLE |
|---|---|
| Joint Base Lewis-McChord<br>Washington | Barracks |

| 5. PROGRAM ELEMENT | 6. CATEGORY CODE | 7. PROJECT NUMBER | 8. PROJECT COST ($000) |
|---|---|---|---|
| 22096A | 72111 | 97456 | Approp 49,000 |

9. COST ESTIMATES

| ITEM | UM (M/E) | QUANTITY | UNIT COST | COST ($000) |
|---|---|---|---|---|
| PRIMARY FACILITY | | | | 34,924 |
| 72111 Barracks | m2 (SF) | 5,572 ( 59,975) | 5,969 | (33,260) |
| 00000 Cyber Security Measures | LS | -- | -- | (500) |
| Sustainability/Energy Measures | LS | -- | -- | (665) |
| Antiterrorism Measures | LS | -- | -- | (499) |
| SUPPORTING FACILITIES | | | | 8,939 |
| Electric Service | LS | -- | -- | (2,829) |
| Water, Sewer, Gas | LS | -- | -- | (321) |
| Steam/Chilled Water Distribution | LS | -- | -- | (39) |
| Paving, Walks, Curbs And Gutters | LS | -- | -- | (2,183) |
| Storm Drainage | LS | -- | -- | (557) |
| Site Imp(2,897) Demo( ) | LS | -- | -- | (2,897) |
| Information Systems | LS | -- | -- | (113) |
| ESTIMATED CONTRACT COST | | | | 43,863 |
| CONTINGENCY (5.00%) | | | | 2,193 |
| SUBTOTAL | | | | 46,056 |
| SUPV, INSP & OVERHEAD (5.70%) | | | | 2,625 |
| TOTAL REQUEST | | | | 48,681 |
| TOTAL REQUEST (ROUNDED) | | | | 49,000 |
| INSTALLED EQT-OTHER APPROP | | | | (0) |

10. Description of Proposed Construction Construct a barracks to accommodate 150 Soldiers. Primary facilities include living and sleeping quarters, baths, storage, service areas, information systems, fire protection and alarm systems, Intrusion Detection System (IDS) installation, and Energy Monitoring Control Systems (EMCS) connection. Supporting facilities include site development, utilities and connections, lighting, paving, parking, walks, curbs and gutters, storm drainage, information systems, landscaping, and signage. Antiterrorism Measures for this project are unique in nature and not included in the unit cost of the building. Heating and air conditioning will be provided by self-contained systems. Measures in accordance with the Department of Defense (DoD) Minimum Antiterrorism for Buildings standards will be provided. Comprehensive building and furnishings-related interior design services are required. Access for individuals with disabilities will be provided. Cyber Security Measures will be incorporated into this project. Sustainability/Energy measures will be provided. Facilities will be designed to a minimum life of 40 years in accordance with DoD's Unified Facilities Criteria (UFC 1-200-02) including energy efficiencies, building envelope and integrated building systems performance. Air Conditioning (Estimated 211 kWr/60 Tons).

11. REQ: 10,239 PN ADQT: 9,533 PN SUBSTD: 2,268 PN

PROJECT: Construct a Barracks at Joint Base Lewis-McChord, Washington. (Current Mission)

REQUIREMENT: This project is required to provide adequate barracks for Soldiers stationed at Joint Base Lewis-McChord (JBLM). This project is required to meet current

DD FORM 1391, JUL 1999 PREVIOUS EDITION IS OBSOLETE PAGE NO. 57

Source: DoD Comptroller's Office, Department of the Army FY2022 Military Construction Program Budget Documents.

**Figure B-1.** Example of DD form 1391: joint base Lewis-McChord barracks (1 of 3). copy of budget justification document submitted to congress for FY2022.

According to the Army, the project “is needed to provide adequate barracks for Soldiers stationed at Joint Base Lewis-McCord (JBLM). This project is required to meet current Army standards to provide a safe and secure environment that provides privacy and comfort. Quality barracks are essential to resilience and quality of life for Soldiers, and an important factor to preserve and enhance the all-volunteer force.”

| 1. COMPONENT<br>Army | FY 2023 MILITARY CONSTRUCTION PROJECT DATA | | 2. DATE<br>01 APR 2022 |
|---|---|---|---|
| 3. INSTALLATION AND LOCATION<br>Joint Base Lewis-McChord<br>Washington | | 4. PROJECT TITLE<br>Barracks | |
| 5. PROGRAM ELEMENT<br>22096A | 6. CATEGORY CODE<br>72111 | 7. PROJECT NUMBER<br>97456 | 8. PROJECT COST ($000)<br>Approp 49,000 |

REQUIREMENT: (CONTINUED)
Army standards to provide a safe and secure environment that provides privacy and comfort. Quality barracks are essential to resilience and quality of life for Soldiers, and an important factor to preserve and enhance the all-volunteer force.
CURRENT SITUATION: Currently, JBLM lacks more than 700 barracks spaces. As a result, the installation is issuing Certificates of Non-Availability (CNAs) and housing allowances to house Soldiers off-base until adequate facilities are available. The lack of proper housing for Soldiers contributes to decreased morale.
IMPACT IF NOT PROVIDED: If this project is not provided, mission readiness and Soldier resilience will degrade, and the need will continue to house Soldiers off-base. Soldiers living off-base have limited access to base support facilities, quality nutrition from dining facilities, and overall increased exposure to off-base risks. Also, lack of adequate barracks near unit compounds reduces Quality-of-Life, morale, equitable support, unit effectiveness, and is expected to degrade retention and Army readiness.
ADDITIONAL: Utility connections are required to privatized electric, water, and wastewater systems. The Army intends to have the electric, water, and wastewater Utilities Privatization (UP) System Owners make and own the necessary connections up to the facility service disconnect or other defined point of demarcation. In the event of a future UP action, the Army intends to have the System Owner make and own the necessary connection supply to the facility disconnect or other defined point of demarcation. Required assessments have been made for supporting facilities and the project is not in a 100-year floodplain in-accordance-with Executive Order 11988. This project has been coordinated with the installation physical security plan, and all physical security measures are included. All required antiterrorism protection measures are included. Alternative methods of meeting this requirement have been explored during project development. This project is the only feasible option to meet the requirement. The Deputy Assistant Secretary of the Army (Installations, Housing and Partnerships) certifies that this project has been considered for joint use potential. The facility will be available for use by other components. Sustainable principles, to include life cycle cost effective practices, will be integrated into the design, development and construction of the project and will follow the guidance detailed in the Army Sustainable Design and Development Policy - complying with applicable laws and executive orders.

12. SUPPLEMENTAL DATA:
A. Estimated Design Data:
(1) Status:
(a) Date Design Started.............................. JUL 2020
(b) Percent Complete as of January 2022.............. 35.00
(c) Date 35% Designed................................ JUN 2021
(d) Date Design Complete............................. JUL 2022
(e) Parametric Cost Estimating Used to Develop Costs.. NO
(f) Type of Design Contract: Design-bid-build

(2) Basis:
(a) Standard or Definitive Design: YES
(b) Where Design Was Most Recently Used:

PAGE NO. 58 PREVIOUS EDITION IS OBSOLETE DD FORM 1391C, JUL 1999

Source: DoD Comptroller’s Office, Department of the Army FY2022 Military Construction Program Budget Documents.

**Figure B-2.** Example of DD Form 1391: Joint Base Lewis-McChord Barracks (1 of 3).

The complete DD Form 1391 for this MILCON project consists of three pages and includes cost estimates, a description of the proposed construction, the Army's requirements and justifications for the project, and other supplemental data. The bottom of the document includes details about the project's timeline – when the design began, the current status of the design process, and projected dates for contract award, start date and completion date.

| 1. COMPONENT | FY 2023 MILITARY CONSTRUCTION PROJECT DATA | | 2. DATE |
|---|---|---|---|
| Army | | | 01 APR 2022 |
| 3. INSTALLATION AND LOCATION | | 4. PROJECT TITLE | |
| Joint Base Lewis-McChord<br>Washington | | Barracks | |

| 5. PROGRAM ELEMENT | 6. CATEGORY CODE | 7. PROJECT NUMBER | 8. PROJECT COST ($000) |
|---|---|---|---|
| 22096A | 72111 | 97456 | Approp 49,000 |

12. SUPPLEMENTAL DATA (CONTINUED..)

A. Estimated Design Data: (CONTINUED..)

Fort George G Meade

| | | |
|---|---|---|
| (c) | Percentage of Design utilizing Standard Design ... | 100 |
| (3) | Total Design Cost (c) = (a)+(b) OR (d)+(e): | ($000) |
| (a) | Production of Plans and Specifications............ | 2,695 |
| (b) | All Other Design Costs............................ | 674 |
| (c) | Total Design Cost................................. | 3,369 |
| (d) | Contract.......................................... | 2,695 |
| (e) | In-house.......................................... | 674 |
| (4) | Construction Contract Award....................... | JAN 2023 |
| (5) | Construction Start................................ | MAR 2023 |
| (6) | Construction Completion........................... | NOV 2024 |

B. Equipment associated with this project which will be provided from other appropriations:

| Equipment Nomenclature | Procuring Appropriation | Fiscal Year Appropriated Or Requested | Cost ($000) |
|---|---|---|---|
| | NA | | |

Installation Engineer: Phone Number: 253-967-3191

DD FORM 1391C, JUL 1999 PREVIOUS EDITION IS OBSOLETE PAGE NO. 59

Source: DoD Comptroller's Office, Department of the Army FY2022 Military Construction Program Budget Documents.

**Figure B-3.** Example of DD Form 1391: Joint Base Lewis-McChord Barracks (1 of 3). Copy of budget justification document submitted to Congress for FY2022.

Chapter 2

# Military Construction, Veterans Affairs, and Related Agencies Appropriation Bill, 2025*

## United States Senate

The Committee on Appropriations reports the bill (S. 4677) making appropriations for military construction, the Department of Veterans Affairs, and related agencies for the fiscal year ending September 30, 2025, and for other purposes, reports favorably thereon and recommends that the bill do pass.

| | |
|---|---|
| *Amounts in new budget authority* | |
| Total of bill as reported to the Senate | $404,527,860,000 |
| Amount of 2024 appropriations | 351,705,297,000 |
| Amount of 2025 budget estimate | 425,483,926,000 |
| Bill as recommended to Senate compared to— | |
| 2024 appropriations | ∂52,822,563,000 |
| 2025 budget estimate | ¥20,956,066,000 |

## Title I: Military Construction

### Hearing

The Subcommittee on Military Construction, Veterans Affairs, and Related Agencies held one hearing related to the fiscal year 2025 military construction

* This is an edited, reformatted and augmented version of Report 118–191, Calendar No. 435, dated July 10,2024.

In: Military Construction and Appropriations
Editor: George K. Santiago
ISBN: 979-8-89530-799-1

budget request. Witnesses included representatives of the Army, Navy, Marine Corps, Air Force, Space Force, and the Office of the Secretary of Defense.

## Summary of Committee Recommendations

The fiscal year 2025 budget request for military construction and family housing totals $17,545,079,000. The Committee recommends $19,307,000,000, which is $1,761,921,000 above the President's budget request.

## Reprogramming Guidelines

The following reprogramming guidelines apply for all military construction and family housing projects. A project or account (including the sub-elements of an account) which has been specifically reduced by the Congress in acting on the budget request is considered to be a congressional interest item and as such, prior approval is required. Accordingly, no reprogrammings to an item specifically reduced below the threshold by the Congress are permitted.

Recognizing the increased cost and complexity of military construction projects, the Committee supports increasing the re-programming threshold to enable the Department to be more agile and reduce delays associated with cost overruns. The reprogramming criteria that apply to military construction projects, as well as new housing construction projects and improvements, is $10,000,000 or 25 percent of the funded amount, whichever is less. To provide the Services the flexibility to proceed with construction contracts without disruption or delay, the costs associated with environmental hazard remediation such as asbestos removal, radon abatement, lead-based paint removal or abatement, and any other legislated environmental hazard remediation may be excluded, provided that such remediation requirements could not be reasonably anticipated at the time of the budget submission. This exclusion applies to projects authorized in this budget year, as well as projects authorized in prior years for which construction has not been completed.

Furthermore, in instances where prior approval of a reprogramming request for a project or account has been received from the Committee, the adjusted amount approved becomes the new base for any future increase or decrease via below-threshold reprogrammings (provided that the project or account is not a congressional interest item as defined above).

In addition to these guidelines, the Services are directed to adhere to the guidance for military construction reprogrammings and notifications, including the pertinent statutory authorities contained in Department of Defense [DoD] Financial Management Regulation 7000.14–R and relevant updates and policy memoranda.

*Facility Sustainment, Restoration, and Modernization [FSRM].*— The Department of Defense is directed to continue describing on form 1390 the backlog of FSRM requirements at installations with future construction projects. For troop housing requests, form 1391 should describe any FSRM conducted in the past 2 years. Likewise, future requirements for unaccompanied housing at the corresponding installation should be included. Additionally, the forms should include English equivalent measurements for projects presented in metric measurement. Rules for funding repairs of facilities under the Operation and Maintenance accounts are described below:

1) components of the facility may be repaired by replacement. Such replacement can be up to current standards or codes;
2) interior arrangements and restorations may be included as repair;
3) additions and new facilities may be done concurrently with repair projects, as long as the final conjunctively funded project is a complete and usable facility; and
4) the appropriate Service Secretary shall notify the appropriate committees prior to carrying out any repair project with an estimated cost in excess of $7,500,000.

*Design Funding.*—The Committee acknowledges that the Department has clarified what construction activities fall under advanced planning, which is funded through the Operation and Maintenance accounts, and which are tied to facility design, which is funded under ''planning and design'' in the Military Construction accounts. To more appropriately reflect DoD activities being appropriated for within this bill, the Committee will now refer to ''design'' funding instead of ''planning and design.''

*Strategic Construction in the Indo-Pacific Region.*—The Committee continues to support efforts to execute emergent requirements and accelerate construction investments in the Indo-Pacific Command [INDOPACOM] region and has provided additional resources above the President's budget request over the past 5 years for this purpose. As such, the Committee provides an additional $53,000,000 for design and authorized major construction that support INDOPACOM objectives in the region.

*Military Installation Resilience.*—Since fiscal year 2020, the Committee has provided dedicated design and unspecified minor construction funding to support military installation resilience. The Committee encourages the services to continue to develop their Installation Climate Resilience Plans and to utilize these documents and identify associated infrastructure requirements that can be addressed through this funding. The Committee believes that such investments are critical to installation readiness and therefore pro vides $30,000,000 in Section 128 to continue to develop and execute projects that will directly enhance military installation resilience. Child Development Centers [CDCs].—The Committee continues to recognize the importance of access to childcare in improving both readiness and quality of life for servicemembers and their families. In order to help address the capacity shortfall and construct modern facilities, in addition to providing $133,000,000 for the CDC projects included in the President's request, the bill provides an additional $31,600,000 in design and major construction funding for three additional projects.

*Defense of Guam Infrastructure.*—The Committee supports the overarching mission of the Enhanced Integrated Air & Missile Defense [EIAMD] system on Guam and recognizes the significant associated environmental, planning, and construction activities that must be undertaken to stand up the program. However, the Committee is concerned about the near-term executability of the full level of funding requested and provides a level which it believes is more realistic given the current progress of the program, allocating excess funding to the Missile Defense Agency [MDA] design account. The Committee strongly encourages MDA, the Departments of the Army and Navy, and other DoD stakeholders to move expeditiously to meet National Environmental Policy Act requirements. Furthermore, in addition to beddown of the missile system, the EIAMD program requires investments in support infrastructure for the personnel and equipment that will be operating it. The Committee notes with concern that such base operations construction is not scheduled to be requested until fiscal year 2026 at the earliest, which will delay personnel beddown or require the use of facilities that are not mission-capable or are improperly configured to support EIAMD.

*Military Construction Funding Initiatives.*—The bill includes $549,421,000 in Section 124 to address important unfunded priorities and cost to completes included in DoD's unfunded priority lists provided to Congress.

*Congressionally Directed Spending.*—The Committee includes Congressionally Directed Spending for design, unspecified minor construction, and major construction in accordance with Senate Rule XLIV. DoD is directed to carry out funding for projects as identified in tables under each relevant account header.

### Military Construction Overview

| | |
|---|---|
| Appropriations, 2024 | $15,690,070,000 |
| Budget estimate, 2025 | 15,561,215,000 |
| Committee recommendation | 16,615,215,000 |

### Military Construction Accounts—Program Description

The military construction appropriation provides for acquisition, construction, installation, and equipment of temporary or permanent public works, military installations, facilities, and real property for the Department of Defense. This appropriation also funds infrastructure projects and programs required to support bases and installations around the world.

### Military Construction, Army

| | |
|---|---|
| Appropriations, 2024 | $2,022,775,000 |
| Budget estimate, 2025 | 2,311,157,000 |
| Committee recommendation | 2,380,477,000 |

### Committee Recommendation

The Committee recommends $2,380,477,000 for Military Construction, Army for fiscal year 2025. This amount is $357,702,000 above the fiscal year 2024 enacted level and $69,320,000 above the budget request. Within the total for Military Construction, Army, $186,820,000 is for the following projects in the following amounts:

## Military Construction, Army

[In thousands of dollars]

| State | Location | Project | Amount |
|---|---|---|---|
| Alabama | Fort Novosel | Crash/Rescue Station at Hatch Army Heliport: Minor Construction. | 6,500 |
| Alabama | Fort Novosel | Crash/Rescue Station at Tabernacle Stagefield: Minor Construction. | 6,600 |
| Alabama | Fort Novosel | Flat Iron Support Facility: Minor Construction | 5,500 |
| Alabama | Fort Novosel | Flight Control Tower—Hanchey Army Airfield: Design. | 1,300 |
| Alabama | Fort Novosel | Installation Service Node Facility: Design | 2,600 |
| Alaska | Fort Wainwright | Dining Facility: Design | 7,700 |
| Alaska | Fort Wainwright | Tactical Equipment Maintenance Facility: Design | 10,800 |
| Arizona | Yuma Proving Ground | Radar Operations & Maintenance Facility: Minor Construction. | 6,000 |
| Georgia | Fort Stewart | Enlisted Unaccompanied Personnel Barracks (PN96725): Design. | 10,500 |
| Georgia | Fort Stewart/Hunter Army Airfield. | SOF Military Working Dog Kennel Facility: Design | 1,230 |
| Hawaii | Pohakuloa Training Area | Airfield Operations Building | 20,000 |
| Illinois | Rock Island Arsenal | Forging Equipment Annex: Minor Construction | 5,000 |
| Illinois | Rock Island Arsenal | New Weapons Quality Assurance/Calibration Facility: Minor Construction. | 4,250 |
| Kansas | Fort Leavenworth | B95 Fire Station Truck Bay Addition: Minor Construction. | 8,200 |
| Kansas | Fort Leavenworth | Fire Station Headquarters: Design | 1,800 |
| Kansas | Fort Leavenworth | Functional Fitness Center: Design | 1,700 |
| Kentucky | Blue Grass Army Depot | Military Container Production Facility: Design | 27,000 |
| Kentucky | Fort Campbell | Foundry Training Facility: Minor Construction | 8,100 |
| Kentucky | Fort Campbell | Modernized Hangar: Design | 11,000 |
| Louisiana | Fort Johnson | Rotational Unit Billeting Area: Design | 9,900 |
| Mississippi | Army Engineer Research and Development Center. | Jet Engine Thermal Simulator: Design | 700 |
| New Mexico | White Sands Missile Range | North Range Mission Control Center: Design | 3,330 |
| New York | Fort Drum | Aircraft Maintenance Hangar Addition: Design | 9,300 |

| State | Location | Project | Amount |
|---|---|---|---|
| New York | Fort Drum | Field Artillery Vehicle Storage Sheds: Design | 800 |
| New York | Fort Drum | | 2,760 |
| New York | Fort Drum | ORTC Transient Training Barracks: Design | 7,790 |
| Washington | Yakima Training Center | Wastewater Treatment Plant: Design | 6,460 |

Further detail of the Committee's recommendation is provided in the State table at the end of this report.

*Air Traffic Control Towers.*—The Committee continues to recognize the importance of Army air traffic control towers, which are the nerve center for an air installation complex and are critical to operational success. However, some current Army towers are antiquated and unsafe. The Committee encourages the Army to allo cate sufficient funding for these critical facilities in future budget requests.

*White Sands Missile Range Energy Infrastructure.*—The Committee acknowledges that power outages at White Sands Missile Range [WSMR] lead to training delays and impact range mission readiness. The Committee encourages U.S. Army Headquarters to work with WSMR to find a mutually agreeable solution to the energy infrastructure needs on the installation. The Committee also directs the U.S. Army to provide a report on mutually agreed upon solution(s) to the energy infrastructure needs of WSMR no later than 120 days after the enactment of this act.

## Military Construction, Navy and Marine Corps

| | |
|---|---|
| Appropriations, 2024 | $5,531,369,000 |
| Budget estimate, 2025 | 4,540,899,000 |
| Committee recommendation | 4,874,699,000 |

## Committee Recommendation

The Committee recommends $4,874,699,000 for Military Construction, Navy and Marine Corps for fiscal year 2025. This amount is $333,800,000 above the budget request. Within the total for Military Construction, Navy and

Marine Corps, $333,800,000 is for the following projects in the following amounts:

Further detail of the Committee's recommendation is provided in the State table at the end of this report.

*Naval Shipyard Modernization.*—The Navy's four public shipyards perform a vital role in national defense by executing maintenance on submarines and aircraft carriers in order to provide combat-ready ships to the fleet. The Committee supports the Department's Shipyard Infrastructure Optimization Plan [SIOP] and believes the restoration of deferred maintenance availabilities along with the public shipyard dry dock investments identified under the plan are critical. The Committee urges the Secretary of the Navy to prioritize the timely funding of public shipyard infrastructure, in particular, the dry dock and shore infrastructure necessary to sup port critical maintenance of surface and submarine fleets. Further, as SIOP continues to refine requirements and synchronize construction activities, the Committee encourages the Department of the Navy to make timely decisions to ensure future maintenance schedules are not interrupted. In particular, as the Navy considers the way forward on the Multi-Mission Dry Dock at Puget Sound Naval Shipyard, the Committee encourages the Department to consider opportunities to pull forward elements of the project which can be responsibly separated out into complete and usable facilities.

Military Construction, Navy and Marine Corps

[In thousands of dollars]

| State | Location | Project | Amount |
|---|---|---|---|
| Arizona | Marine Corps Air Station Yuma. | UDP Transient Barracks: Design | 3,800 |
| Arizona | Marine Corps Air Station Yuma. | Water Treatment Plant | 90,000 |
| Connecticut | Naval Submarine Base New London. | Submarine Storage, Maintenance & Operations Facility: Design. | 3,200 |
| Hawaii | Joint Base Pearl Harbor-Hickam. | (Red Hill) Water Treatment Plant | 90,000 |
| Maine | Portsmouth Naval Shipyard | Parking Structure: Design | 9,500 |
| Maryland | Naval Surface Warfare Center Indian Head. | Contained Burn Facility | 50,000 |
| South Carolina | Marine Corps Air Station Beaufort. | Aircraft Maintenance Hangar: Cost to Complete | 61,700 |
| Virginia | Marine Corps Base Quantico | Child Development Center: Design | 8,200 |
| Virginia | Naval Air Station Oceana | Unaccompanied Housing: Design | 16,000 |
| Virginia | Norfolk Naval Shipyard | Child Development Center: Design | 1,400 |

## Military Construction, Air Force

| | |
|---|---|
| Appropriations, 2024 | $2,741,424,000 |
| Budget estimate, 2025 | 3,187,126,000 |
| Committee recommendation | 3,549,626,000 |

## Committee Recommendation

The Committee recommends $3,549,626,000 for Military Construction, Air Force for fiscal year 2025. This amount is $808,202,000 above the fiscal year 2024 enacted level and $362,500,000 above the budget request. Within the total for Military Construction, Air Force, $366,000,000 is for the following projects in the following amounts:

Military Construction, Air Force

[In thousands of dollars]

| State | Location | Project | Amount |
|---|---|---|---|
| Alaska | Eielson Air Force Base. | Joint Mobility Center Expansion: Design | 5,800 |
| Alaska | Eielson Air Force Base | Permanent Party Dormitory: Design | 6,700 |
| Alaska | Joint Base Elmendorf-Richardson. | Precision Guided Munitions Complex: Design | 15,600 |
| California | Beale Air Force Base | Multi-Domain Operations Complex | 44,000 |
| Colorado | Buckley Space Force Base | Power Independence | 60,000 |
| Hawaii | Joint Base Pearl Harbor-Hickam. | General Purpose Maintenance Hangar: Design | 7,000 |
| Hawaii | Joint Base Pearl Harbor-Hickam. | Mobility Processing Center/War Reserve Material Warehouse: Design. | 11,000 |
| Louisiana | Barksdale Air Force Base | Child Development Center | 22,000 |
| Mississippi | Keesler Air Force Base | Air Traffic Control Tower | 25,000 |
| Nebraska | Offutt Air Force Base | Consolidated Training Complex/Professional Development Center: Design. | 6,000 |
| Nebraska | Offutt Air Force Base | Survivable Airborne Operations Center Complex: Design. | 70,000 |
| Nevada | Nellis Air Force Base | Regional Communications Center: Design | 3,500 |

Military Construction, Air Force (Continued)

| State | Location | Project | Amount |
|---|---|---|---|
| New Mexico | Cannon Air Force Base | ADAL Security Forces Facility: Minor Construction | 4,000 |
| New Mexico | Cannon Air Force Base | Special Operations Forces DAGRE Facility: Minor Construction. | 7,200 |
| New Mexico | Holloman Air Force Base | Holloman High-Speed Test Track: Design | 22,400 |
| North Carolina | Seymour Johnson Air Force Base. | Combat Arms Training and Maintenance Complex | 15,000 |
| Ohio | Wright-Patterson Air Force Base. | Advanced Materials Research Laboratory—C2A: Design. | 9,200 |
| Ohio | Wright-Patterson Air Force Base. | Dormitory: Design | 3,400 |
| Ohio | Wright-Patterson Air Force Base. | U.S. Space Force Intelligence Center: Design | 19,000 |
| Oklahoma | Altus Air Force Base | Main Gate Entry Control Facility: Design | 2,200 |
| Washington | Fairchild Air Force Base | KC–135 ADAL Aircraft Parts Warehouse: Design | 2,500 |
| Washington | Fairchild Air Force Base | Taxilane Pull-Through Spots: Design | 4,500 |

Further detail of the Committee's recommendation is provided in the State table at the end of this report.

*Air Force Special Operations Command Power Projection Infrastructure.*—In July 2023 the Air Force informed the Committee of its decision to establish the 492nd Special Operations Wing at Davis-Monthan Air Force Base as an Air Force Special Operations Command Power Projection Wing, which requires infrastructure investments as more than a half dozen squadrons will be relocated to and activated at the installation. The Committee expects this plan to be carried out as communicated and encourages the Air Force, in collaboration with Special Operations Command, to work expeditiously to identify and program military construction requirements in order to meet established beddown and operational timelines. Furthermore, the Committee directs the Air Force to immediately communicate any deviations from the final decision with regards to composition of the Wing or capability delays, and to brief the Committees on Appropriations of both Houses of Congress no later than 90 days after enactment of this act on any requirements identified after submission of the fiscal year 2025 budget request for operational or base support infrastructure.

*Hypersonic Labs.*—The Committee continues to recognize the importance of Air Force hypersonic test range facilities and that investment is needed in testing infrastructure to support critical hypersonic weapons development. The Committee encourages the Department to make investments in its testing infrastructure, including basic water and energy capacity, turbine propulsion cycles, and laboratory infrastructure, in order to advance existing capabilities and develop a wide range of new capabilities to compete with advanced adversaries.

*Global Strike Command Entry Control Points.*—The Committee encourages Air Force Global Strike Command to leverage minor military construction funding to enhance efficiency and decrease operational risk at the entry points to weapons storage areas.

*Long Range Stand Off Infrastructure.*—The Committee notes that Air Force efforts to construct weapons generation facilities [WGF] have incurred significant cost and schedule delays and is concerned that similar challenges could delay the entry into service of the Long Range Stand Off [LRSO] weapon. Therefore, in addition to the WGF quarterly status updates, the Committee directs the Secretary of the Air Force to provide a report within 120 days of enactment of this act that outlines a construction schedule that ensures at least two domestic bases have the facilities necessary to operate and maintain the LRSO when the Air Force takes delivery of the weapons system.

*B–21 Infrastructure.*—The Committee recognizes the importance of environmental protection shelters that help protect aircraft from extreme weather but is concerned about the suitability of open-sided shelters for B–21 bomber aircraft operating out of installations that are at-risk of severe weather events. The Committee supports developing and incorporating design elements to address severe weather resiliency, specifically to protect the safety of airmen, protect the lifespan of the B–21 aircraft and its low-observable coating, and promote high mission capable rates. The Committee encourages the Air Force to consider additional elements beyond overhead protection that could be used in environmental protection shelters to protect aircraft and airmen in extreme weather events. *Spaceport Infrastructure.* —The Committee recognizes the importance of investing in spaceport infrastructure in order to accommodate an increase in national security launches in the coming years and to ensure that such infrastructure is maintained, updated, and expanded, where necessary. Therefore, the Committee directs the Space Force to submit a report no later than 120 days after enactment of this act identifying spaceport infrastructure shortfalls, including an identification of military construction projects that are deemed essential to accommodating increased launch cadence, and potential avenues to address identified shortfalls.

*Air Force Civil Engineer Center.*—The Committee is aware that the Air Force is evaluating options to improve engineering supervision and oversight, and budgeting, related to overseas projects where it is the designated DoD Construction Agent. The Committee is supportive of this endeavor in these limited circumstances and encourages the Air Force to communicate any proposals to the Committees on Appropriations of both Houses of Congress.

*Child Development Centers.*—The Air Force has numerous unfunded child development centers that are at 35 percent design but have not been prioritized for inclusion across the Future Years Defense Program [FYDP]. Recognizing that the FYDP is a working document which is updated annually, the Committee is concerned that the lack of inclusion of these and related projects in prior FYDPs has a negative impact on the ability of the Air Force to fund new childcare facilities. Therefore, the Committee encourages the Air Force to better incorporate child development center programming into its future FYDPs.

*Sentinel Infrastructure Costs.*—The Committee is concerned that, despite significant increases in the estimated costs to replace launch facility [LF] infrastructure to support the Sentinel program, the Air Force has not fully assessed the condition of each Minuteman III LF, and therefore directs the Secretary of the Air Force to provide a briefing on the service's efforts to assess the condition of existing Minuteman III LF infrastructure, including the number of LFs it has assessed to date, plans to conduct additional LF assessments as part of the restructuring of the Sentinel program, and the estimated cost of completing assessments of all 450 LFs.

## Military Construction, Defense-Wide

(INCLUDING TRANSFER OF FUNDS)

| | |
|---|---|
| Appropriations, 2024 | $3,161,782,000 |
| Budget estimate, 2025 | 3,733,163,000 |
| Committee recommendation | 3,601,163,000 |

## Committee Recommendation

The Committee recommends $3,601,163,000 for projects considered within the Military Construction, Defense-Wide account for fiscal year 2025. This amount is $439,381,000 above the fiscal year 2024 enacted level. Within the total for Military Construction, Defense-Wide, $4,000,000 is for the following projects in the following amounts:

Military Construction, Defense-Wide

[In thousands of dollars]

| State | Location | Project | Amount |
|---|---|---|---|
| Mississippi ... | Key Field .................... | Upgrade Fuel Hydrant System: Design | 1,000 |
| Oregon ......... | Klamath Falls Airport— Kingsley Field. | Fuel Facility: Design . | 3,000 |

Further detail of the Committee's recommendation is provided in the State table at the end of this report.

*Linear Generators.*—The Committee is aware of linear generator technology that could improve grid resilience and energy efficiency. The Committee encourages the Department to explore the potential applicability of this technology and directs the Secretary of Defense to submit a report no later than 120 days after enactment of this act detailing the potential use of linear generators on military installations, limitations to incorporating this technology into existing or new energy systems, and any other relevant information.

*Innovative Construction Materials.*—The Joint Explanatory Statement accompanying Public Law 117–328 included a reporting requirement directing the Assistant Secretary of Defense for Energy, Installations, and Environment to provide a report to the Committees on its efforts to incorporate sustainable products into the Unified Facilities Criteria and Unified Facilities Guide Specifications, integrate these materials into project designs, and resources needed to accomplish these efforts. The Committee eagerly awaits delivery of that report.

*Joint Spectrum Center.*—The Committee is encouraged by the plans proposed by the DoD to relocate the Joint Spectrum Center from its existing location in Annapolis, Maryland to a modern and secure facility on Fort Meade. The Committee urges the Department to provide a plan to allocate the necessary resources to complete the relocation as soon as possible.

## Military Construction, Army National Guard

| | |
|---|---|
| Appropriations, 2024 | $620,647,000 |
| Budget estimate, 2025 | 362,129,000 |
| Committee recommendation | 456,459,000 |

## Committee Recommendation

The Committee recommends $456,459,000 for Military Construction, Army National Guard for fiscal year 2025. This amount is $94,330,000 above the budget request. Within the total for Military Construction, Army National Guard, $101,330,000 is for the following projects in the following amounts:

Military Construction, Army National Guard

[In thousands of dollars]

| State | Location | Project | Amount |
|---|---|---|---|
| Connecticut ...... | Groton .................... | Aircraft Maintenance Hangar: | 6,500 |
| Delaware .......... | Dagsboro ................. | Design Dagsboro Readiness Center: | 7,500 |
| Illinois .............. | Marseilles Training Center | Minor Construction Marseilles Training Center Range | 4,300 |
| Maine ............. | Saco ..................... | Renovation: Minor Construction. Southern Maine Readiness Center: Design | 800 |
| Maine ............... | Woodville .................. | Automated Qualification Training Range: Design .... | 1,000 |
| Maryland .......... | Hagerstown .......................... | Field Maintenance Shop: Design .............................. | 2,470 |
| Michigan .......... | Olympia Armory .................... | Readiness Center Addition/Alteration: Design ......... | 4,400 |
| Nebraska ......... | Hastings ................ | Collective Training Unaccompanied Housing: Minor Construction. | 5,900 |
| New Hampshire | Concord .................... | Lakota Hangar Addition and Environmental Upgrades: Design. | 1,200 |
| New Hampshire | Pembrooke ................ | Readiness Center Addition/Alteration: Minor Construction. | 6,500 |
| Oregon ............ | Naval Weapons Systems Training Facility Boardman. | Multi-Purpose Machine Gun Range: Design .. | 1,690 |
| Pennsylvania ... | Fort Indiantown Gap Training Site. | Ready Building: Design ................ | 3,250 |

| State | Location | Project | Amount |
|---|---|---|---|
| Rhode Island .... | North Kingstown ...... | National Guard Readiness Center: Cost to Complete. | 16,000 |
| Vermont .......... | St. Albans ................ | Readiness Center: Design ............ | 4,030 |
| Washington ..... | Tacoma .................... | Controlled Humidity Warehouse: Minor Construction | 6,000 |
| West Virginia ... | Bluefield ................... | National Guard Readiness Center: Design .............. | 4,000 |
| West Virginia ... | Camp Dawson ......... | Fire Station Support Building: Minor Construction .. | 6,000 |
| West Virginia ... | Camp Dawson-Kingwood ..... | General Instruction Building: Design ...................... | 1,200 |
| West Virginia ... | Joint Forces Headquarters—Charleston. | National Guard/Reserve Center Building: Design .... | 1,400 |
| West Virginia ... | Joint Forces Headquarters—Charleston. | Physical Fitness Center (Multipurpose Building): Design. | 2,500 |
| Wisconsin ....... | Black River Falls ................ | National Guard Readiness Center: Design .............. | 2,600 |
| Wisconsin ........ | Wausau ............................... | National Guard Readiness Center: Design .............. | 7,150 |
| Wisconsin ........ | Wisconsin Rapids ..... | National Guard Readiness Center: Design .............. | 4,940 |

Further detail of the Committee's recommendation is provided in the State table at the end of this report.

*Lodging Facilities.*—The Committee continues to highlight lodging shortfalls near Army National Guard training centers that support training and readiness activities as well as support training for outside agencies. The Committee encourages the Army National Guard to allocate appropriate funding towards lodging requirements in future budget requests.

## Military Construction, Air National Guard

| | |
|---|---|
| Appropriations, 2024 | $295,526,000 |
| Budget estimate, 2025 | 190,792,000 |
| Committee recommendation | 307,792,000 |

## Committee Recommendation

The Committee recommends $307,792,000 for Military Construction, Air National Guard for fiscal year 2025. This amount is $12,266,000 above the fiscal year 2024 enacted level and $117,000,000 above the budget request. Within the total for Military Construction, Air National Guard, $113,000,000 is for the following projects in the following amounts:

Military Construction, Air National Guard

[In thousands of dollars]

| State | Location | Project | Amount |
|---|---|---|---|
| Arizona | Morris Air National Guard Base. | Base Entry Complex: Cost to Complete | 7,000 |
| Delaware | New Castle County Airport | Entry Control Facility: Minor Construction | 8,000 |
| Kentucky | Louisville International Airport-Standiford Fld. | C–130J ADAL Fuel Cell Hangar: Minor Construction | 8,500 |
| Maine | Bangor International Airport | Aircraft Maintenance Squadron/Aircraft Ground Equipment Facility: Design. | 800 |
| Maine | Bangor International Airport | Fuel Cell Hangar | 50,000 |
| Minnesota | Minneapolis St Paul International Airport. | C–130J Fuel Cell Maintenance Hangar: Design | 4,300 |
| Mississippi | Key Field | Fuel Cell/Corrosion Control Hangar: Design | 5,000 |
| Mississippi | Key Field | Maintenance Hangar: Design | 5,000 |
| Nevada | Reno-Tahoe International Airport. | Aircraft Parking Ramp Extension: Design | 1,000 |
| Nevada | Reno-Tahoe International Airport. | Aircraft Support Equipment and Engine Facility: Design. | 2,000 |
| Nevada | Reno-Tahoe International Airport. | Fuel Cell Hangar: Design | 3,000 |
| New Jersey | Atlantic City International Airport. | Fighter Operations Facility: Design | 3,400 |
| Oregon | Klamath Falls Airport—Kingsley Field. | Aircraft Shelter MCCA: Design | 4,300 |
| Pennsylvania | Pittsburgh International Airport. | Entry Control Facility: Minor Construction | 5,000 |
| Washington | Camp Murray | National Guard/Reserve Center (ANG Cost Share): Minor Construction. | 5,700 |

Further detail of the Committee's recommendation is provided in the State table at the end of this report.

## Military Construction, Army Reserve

| | |
|---|---|
| Appropriations, 2024 | $151,076,000 |
| Budget estimate, 2025 | 255,032,000 |
| Committee recommendation | 375,682,000 |

## Committee Recommendation

The Committee recommends $375,682,000 for Military Construction, Army Reserve for fiscal year 2025. This amount is $224,606,000 above the fiscal year 2024 enacted level and $120,650,000 above the budget request. Within the total for Military Construction, Army Reserve, $120,650,000 is for the following projects in the following amounts:

Military Construction, Army Reserve

[In thousands of dollars]

| State | Location | Project | Amount |
|---|---|---|---|
| California ......... | Bell ......................... | Army Reserve Center: Design ........... | 3,750 |
| Georgia ............ | Marine Corps Logistics Base Albany. | Army Reserve Center ........................ | 40,000 |
| Kentucky ......... | Fort Knox ................ | Aviation Support Facility ........................................ | 75,000 |
| Pennsylvania .... | New Castle .............. | Area Maintenance Support Activity and Land: Design. | 1,750 |
| Wisconsin ........ | Manitowoc ............ | Army Reserve Vehicle Maintenance Shop: Design ... | 150 |

Further detail of the Committee's recommendation is provided in the State table at the end of this report.

## Military Construction, Navy Reserve

| | |
|---|---|
| Appropriations, 2024 | $51,291,000 |
| Budget estimate, 2025 | 29,829,000 |
| Committee recommendation | 29,829,000 |

## Committee Recommendation

The Committee recommends $29,829,000 for Military Construction, Navy Reserve for fiscal year 2025. This amount is equal to the budget request. Further detail of the Committee's recommendation is provided in the State table at the end of this report.

### Military Construction, Air Force Reserve

| | |
|---|---|
| Appropriations, 2024 | $331,572,000 |
| Budget estimate, 2025 | 69,263,000 |
| Committee recommendation | 107,663,000 |

### Committee Recommendation

The Committee recommends $107,663,000 for Military Construction, Air Force Reserve for fiscal year 2025. This amount is $38,400,000 above the budget request. Within the total for Military Construction, Air Force Reserve, $38,400,000 is for the following projects in the following amounts:

Military Construction, Air Force Reserve

[In thousands of dollars]

| State | Location | Project | Amount |
|---|---|---|---|
| New York ........ | Niagara Falls Air Reserve Station. | TWY A4 and Runway 28R Overrun: Design .............. | 5,400 |
| South Carolina . | Joint Base Charleston | Aeromedical Evacuation Facility ...... | 33,000 |

Further detail of the Committee's recommendation is provided in the State table at the end of this report.

## North Atlantic Treaty Organization

### Security Investment Program

| | |
|---|---|
| Appropriations, 2024 | $293,434,000 |
| Budget estimate, 2025 | 433,864,000 |
| Committee recommendation | 433,864,000 |

### Program Description

The North Atlantic Treaty Organization [NATO] appropriation provides for the U.S. cost share of the NATO Security Investment Program for the acquisition and construction of military facilities and installations (including international military headquarters) and for related expenses for the collective defense of the NATO Treaty area.

### Committee Recommendation

The Committee recommends $433,864,000 for the NATO Security Investment Program for fiscal year 2025. This amount is equal to the budget request.

### Department of Defense Base Closure Account

| | |
|---|---|
| Appropriations, 2024 | $489,174,000 |
| Budget estimate, 2025 | 447,961,000 |
| Committee recommendation | 497,961,000 |

### Program Description

Section 2711 of the National Defense Authorization Act for Fiscal Year 2014 (Public Law 112–239) consolidated the Base Closure Account 1990 and the Base Closure Account 2005 into a single Department of Defense Base Closure Account. The Base Closure Account provides for cleanup and disposal of property consistent with the four closure rounds required by the base closure acts of 1988 and 1990, and with the 2005 closure round required by the Defense Base Closure and Realignment Act of 1990 (10 U.S.C. 2687 note).

### Committee Recommendation

The Committee recommends a total of $497,961,000 for the Department of Defense Base Closure Account for fiscal year 2025. This amount is $50,000,000 above the budget request.

*Environmental Remediation.*—The Committee continues to recognize that the Services have active costs associated with ordnance disposal and environmental clean-up, to include per- and polyfluoroalkyl substance contamination, at dozens of realigned or closed installations. Therefore, the Committee provides an additional $50,000,000 to help remediate hazardous environmental sites at installations closed under previous Base Closure and Re-alignment rounds. The Committee directs the Secretary of Defense, in coordination with the Service Secretaries, to provide a spend plan for these additional funds to the Committees on Appropriations of both Houses of Congress no later than 60 days after enactment of this act.

**Family Housing Overview**

| | |
|---|---|
| Appropriations, 2024 | $1,970,255,000 |
| Budget estimate, 2025 | 1,983,367,000 |
| Committee recommendation | 1,983,367,000 |

**Family Housing Accounts—Program Description**

The Family Housing appropriation provides funds for military family housing construction activities, operation and maintenance, the Family Housing Improvement Fund, and the Homeowners Assistance Program. Construction accounts provide funding for new construction, improvements and the Federal Government share of housing privatization. Operation and maintenance accounts fund costs associated with the maintenance and leasing of military family housing, including utilities, services, management, and furnishings.

**Committee Recommendation**

The Committee recommends $1,983,367,000 for Family Housing Construction, Operation and Maintenance, and the Department's family housing improvement fund for fiscal year 2025. This amount is equal to the budget request.

**Family Housing Construction, Army**

| | |
|---|---|
| Appropriations, 2024 | $304,895,000 |
| Budget estimate, 2025 | 276,647,000 |

| | |
|---|---|
| Committee recommendation | 276,647,000 |

**Committee Recommendation**

The Committee recommends $276,647,000 for Family Housing Construction, Army for fiscal year 2025. This amount is equal to the budget request. Further detail of the Committee's recommendation is provided in the State table at the end of this report.

**Family Housing Operation and Maintenance, Army**

| | |
|---|---|
| Appropriations, 2024 | $395,485,000 |
| Budget estimate, 2025 | 475,611,000 |
| Committee recommendation | 475,611,000 |

**Committee Recommendation**

The Committee recommends $475,611,000 for Family Housing Operation and Maintenance, Army for fiscal year 2025. This amount is equal to the budget request.

**Family Housing Construction, Navy and Marine Corps**

| | |
|---|---|
| Appropriations, 2024 | $277,142,000 |
| Budget estimate, 2025 | 245,742,000 |
| Committee recommendation | 245,742,000 |

**Committee Recommendation**

The Committee recommends $245,742,000 for Family Housing Construction, Navy and Marine Corps for fiscal year 2025. This amount is equal to the budget request. Further detail of the Committee's recommendation is provided in the State table at the end of this report.

## Family Housing Operation and Maintenance, Navy and Marine Corps

| | |
|---|---|
| Appropriations, 2024 | $373,854,000 |
| Budget estimate, 2025 | 377,217,000 |
| Committee recommendation | 377,217,000 |

### Committee Recommendation

The Committee recommends $377,217,000 for Family Housing Operation and Maintenance, Navy and Marine Corps for fiscal year 2025. This amount is equal to the budget request.

*Naval Station Great Lakes.*—The Committee is concerned about the limited military family housing available at Naval Station [NS] Great Lakes and the resulting reliance on civilian housing, as well as the impact of condemned housing on the safety and security of the local community. The Committee directs the Secretary of the Navy to provide a report no later than 90 days after enactment of this act to the Committees on Appropriations of both Houses of Congress on the Department's plans to address housing capacity and condition shortfalls at NS Great Lakes, to include identifying any associated funding costs for remediation and/or disposal of property, and an anticipated timeline. The report should also include an assessment of the feasibility and impact of a new housing construction project to accommodate accompanied service members.

## Family Housing Construction, Air Force

| | |
|---|---|
| Appropriations, 2024 | $237,097,000 |
| Budget estimate, 2025 | 221,549,000 |
| Committee recommendation | 221,549,000 |

### Committee Recommendation

The Committee recommends $221,549,000 for Family Housing Construction, Air Force for fiscal year 2025. This amount is equal to the budget request.

Further detail of the Committee's recommendation is provided in the State table at the end of this report.

**Family Housing Operation and Maintenance, Air Force**

| | |
|---|---|
| Appropriations, 2024 | $324,386,000 |
| Budget estimate, 2025 | 326,250,000 |
| Committee recommendation | 326,250,000 |

**Committee Recommendation**

The Committee recommends $326,250,000 for Family Housing Operation and Maintenance, Air Force for fiscal year 2025. This amount is equal to the budget request.

**Family Housing Operation and Maintenance, Defense-Wide**

| | |
|---|---|
| Appropriations, 2024 | $50,785,000 |
| Budget estimate, 2025 | 52,156,000 |
| Committee recommendation | 52,156,000 |

**Committee Recommendation**

The Committee recommends $52,156,000 for Family Housing Operation and Maintenance, Defense-Wide for fiscal year 2025. This amount is equal to the budget request.

**Department of Defense Family Housing Improvement Fund**

| | |
|---|---|
| Appropriations, 2024 | $6,611,000 |
| Budget estimate, 2025 | 8,195,000 |
| Committee recommendation | 8,195,000 |

## Program Description

The Family Housing Improvement Fund appropriation provides for the Department of Defense to undertake housing initiatives and to provide an alternative means of acquiring and improving military family housing and supporting facilities. This account provides seed money for housing privatization initiatives.

## Committee Recommendation

The Committee recommends $8,195,000 for the Family Housing Improvement Fund for fiscal year 2025. This amount is equal to the budget request.

## Department of Defense Military Unaccompanied Housing Improvement Fund

| | |
|---|---|
| Appropriations, 2024 | $496,000 |
| Budget estimate, 2025 | 497,000 |
| Committee recommendation | 497,000 |

## Program Description

The Military Unaccompanied Housing Improvement Fund appropriation provides for the Department of Defense to undertake housing initiatives and to provide an alternative means of acquiring and improving military unaccompanied housing and supporting facilities. This account provides seed money for housing privatization initiatives.

## Committee Recommendation

The Committee recommends $497,000 for the Military Unaccompanied Housing Improvement Fund for fiscal year 2025. This amount is equal to the budget request.

## Administrative Provisions

Sec. 101. The Committee includes a provision that restricts payments under a cost-plus-a-fixed-fee contract for work, except in cases of contracts for environmental restoration at base closure sites.

Sec. 102. The Committee includes a provision that permits the use of funds for the hire of passenger motor vehicles.

Sec. 103. The Committee includes a provision that permits the use of funds for defense access roads.

Sec. 104. The Committee includes a provision that prohibits construction of new bases inside the continental United States for which specific appropriations have not been made.

Sec. 105. The Committee includes a provision that limits the use of funds for purchase of land or land easements.

Sec. 106. The Committee includes a provision that prohibits the use of funds to acquire land, prepare a site, or install utilities for any family housing except housing for which funds have been made available.

Sec. 107. The Committee includes a provision that limits the use of minor construction funds to transfer or relocate activities among installations.

Sec. 108. The Committee includes a provision that prohibits the procurement of steel unless American producers, fabricators, and manufacturers have been allowed to compete.

Sec. 109. The Committee includes a provision that prohibits payments of real property taxes in foreign nations.

Sec. 110. The Committee includes a provision that prohibits construction of new bases overseas without prior notification.

Sec. 111. The Committee includes a provision that establishes a threshold for American preference of $500,000 relating to architect and engineering services for overseas projects.

Sec. 112. The Committee includes a provision that establishes preference for American contractors for military construction in the United States territories and possessions in the Pacific, and on Kwajalein Atoll, or in countries bordering the Arabian Gulf.

Sec. 113. The Committee includes a provision that requires notification of military exercises involving construction in excess of $100,000.

Sec. 114. The Committee includes a provision that permits funds appropriated in prior years to be available for construction authorized during the current session of Congress.

Sec. 115. The Committee includes a provision that permits the use of expired or lapsed funds to pay the cost of supervision for any project being completed with lapsed funds.

Sec. 116. The Committee includes a provision that permits obligation of funds from more than one fiscal year to execute a construction project, provided that the total obligation for such project is consistent with the total amount appropriated for the project.

Sec. 117. The Committee includes a provision that permits the transfer of funds from Family Housing Construction accounts to the DoD Family Housing Improvement Fund and from Military Construction accounts to the DoD Military Unaccompanied Housing Improvement Fund.

Sec. 118. The Committee includes a provision that provides transfer authority to the Homeowners Assistance Fund.

Sec. 119. The Committee includes a provision that requires all acts making appropriations for military construction be the sole funding source of all operation and maintenance for family housing, including flag and general officer quarters, and limits the repair on flag and general officer quarters to $35,000 per unit per year without prior notification to the congressional defense committees.

Sec. 120. The Committee includes a provision that provides authority to expend funds from the ''Ford Island Improvement'' account.

Sec. 121. The Committee includes a provision that allows the transfer of expired funds to the Foreign Currency Fluctuation, Construction, Defense Account.

Sec. 122. The Committee includes a provision that allows the reprogramming of military construction and family housing construction funds among projects and activities within the account in which they are funded.

Sec. 123. The Committee includes a provision that prohibits the use of funds in this title for design and construction of projects at Arlington National Cemetery.

Sec. 124. The Committee includes a provision providing additional funds for unfunded military construction priorities.

Sec. 125. The Committee includes a provision directing all amounts appropriated to military construction accounts be immediately available and allotted for the full scope of authorized projects.

Sec. 126. The Committee includes a provision permitting the obligation of funds for fiscal year 2017, fiscal year 2018, fiscal year 2019, and fiscal year

2020 projects for which the project authorization has not lapsed or has been extended.

Sec. 127. The Committee includes a provision defining the congressional defense committees.

Sec. 128. The Committee includes a provision providing additional funding for Military Installation Resilience.

Sec. 129. The Committee includes a provision providing additional funding for design and authorized major construction projects at foreign military training sites.

Sec. 130. The Committee includes a provision prohibiting the use of funds in this act to close or realign Naval Station Guantanamo Bay, Cuba.

Sec. 131. The Committee includes a provision providing funding for unfunded authorized major construction projects.

## Title II. Department of Veterans Affairs

### Hearing

The Subcommittee on Military Construction, Veterans Affairs, and Related Agencies held one hearing related to the Department of Veterans Affairs' budget request for fiscal year 2025 and fiscal year 2026 advance appropriations. The subcommittee heard testimony from the Honorable Denis R. McDonough, Secretary of Veterans Affairs.

### Summary of Committee Recommendations

The Committee recommendation includes $339,591,940,000 for the Department of Veterans Affairs [VA] for fiscal year 2025, including $210,412,794,000 in mandatory spending and $129,179,146,000 in discretionary spending. The Committee also recommends $222,227,506,000 in advance appropriations for mandatory benefits and $131,439,000,000 in advance appropriations for veterans medical care for fiscal year 2026.

## Department Overview

The Veterans Administration was established on July 21, 1930, as an independent agency by Executive Order 5398, in accordance with the act of July 3, 1930 (46 Stat. 1016). This act authorized the President to consolidate and coordinate Federal agencies specially created for or concerned with the administration of laws providing benefits to veterans, including the Veterans' Bureau, the Bureau of Pensions, and the National Home for Disabled Volunteer Soldiers. On March 15, 1989, the Veterans Administration was elevated to Cabinet level status as the Department of Veterans Affairs.

VA's mission is to serve America's veterans and their families as their principal advocate in ensuring they receive the care, support, and recognition they have earned in service to the Nation. As of September 30, 2023, there were an estimated 18.2 million living veterans residing in the United States, American Samoa, Guam, the Northern Mariana Islands, Puerto Rico, and the U.S. Virgin Islands. There were an estimated 21.5 million dependents (spouses and dependent children) of living veterans, and there were about 600,000 survivors of deceased veterans receiving VA survivor benefits. Thus, approximately 40.3 million people, or 11.9 percent of the total estimated resident population of the United States, Puerto Rico, U.S. Island Areas and U.S. citizens living abroad were recipients or potential recipients of veterans benefits from the Federal Government. VA's operating units include the Veterans Benefits Administration, Veterans Health Administration, National Cemetery Administration, and staff support offices.

The Veterans Benefits Administration [VBA] provides an integrated program of nonmedical veterans benefits. VBA administers a broad range of benefits to veterans and other eligible beneficiaries through 4 district offices; 56 regional offices; 3 Pension Management Centers; 6 Fiduciary Hubs; 2 Regional Education Processing Offices; 8 Regional Loan Centers; 1 Insurance Center; and 1 National Contact Center. The benefits provided include compensation for service-connected disabilities; non-service-connected pensions for veterans who meet certain wartime service, income and net worth, age and/or disability requirements; vocational rehabilitation assistance; educational and training assistance; home buying assistance; estate protection services for veterans under legal disability; information and assistance through personalized contacts; and 11 life insurance programs.

The Veterans Health Administration [VHA] develops, maintains, and operates a national healthcare delivery system for eligible veterans; carries out a program of education and training of healthcare personnel; conducts medical research and development; and furnishes health services to members of the Armed Forces during periods of war or national emergency. A system consisting of 18 Veterans Integrated Service Networks [VISN]; 145 VA Hospitals; 12 Health Care Centers; 289 Multi-Specialty Community- Based Outpatient Clinics; 415 Primary Care Community-Based Outpatient Clinics; 415 Outpatient Services Sites; 135 Community Living Centers; 127 Mental Health Residential rehabilitation treatment programs; 300 readjustment counseling Vet Centers; and 83 Mobile Vet Centers is maintained to meet VA's medical mission.

The National Cemetery Administration [NCA] provides for the interment of the remains of eligible deceased servicemembers and discharged veterans in any national cemetery with available grave space; permanently maintains these graves; provides headstones and markers for the graves of eligible persons in national and private cemeteries; administers the grant program for aid to States in establishing, expanding, or improving State veterans cemeteries; and provides certificates to families of deceased veterans recognizing their contributions and service to the Nation. In 2025, cemetery activities will encompass 158 national cemeteries and 35 cemeterial installations and their maintenance as national shrines. Staff support offices include the Office of Inspector General, Board of Veterans Appeals, and General Administration offices, which support the Secretary, Deputy Secretary, Under Secretary for Benefits, Under Secretary for Health, Under Secretary for Memorial Affairs, and General Counsel.

**Veterans Benefits Administration**

| | |
|---|---|
| Appropriations, 2024 | $176,750,703,000 |
| Advance Appropriations, 2025 | 195,845,439,000 |
| Budget estimate, 2025 | 18,928,355,000 |
| Committee recommendation, 2025 | 18,928,355,000 |
| Budget estimate, advance appropriations, 2026. | 222,227,506,000 |
| Committee recommendation, advance appropriations, 2026 | 22,227,506,000 |

## Administration Overview

The Veterans Benefits Administration is responsible for the payment of compensation and pension benefits to eligible service-connected disabled veterans, as well as education benefits and housing loan guarantees.

## Committee Recommendation

In fiscal year 2024, the Committee provided $195,845,439,000 in advance appropriations for VBA for fiscal year 2025. This included $182,310,515,000 for Compensation and Pensions; $13,399,805,000 for Readjustment Benefits; and $135,199,422 for Veterans Insurance and Indemnities.

For fiscal year 2025, the Committee's recommendation includes an additional $9,820,699,000 for Compensations and Pensions; an additional $2,657,656,000 for Readjustment Benefits; $319,596,460 for the Veterans Housing Benefit Program Fund administrative expenses; $64,431 for the Vocational Rehabilitation Loans Program account, with $493,868 for administrative expenses; $5,845,241 for the Native American Veteran Housing Loan Program account; and $4,035,000,000 for the General Operating Expenses, Veterans Benefits Administration account.

The Committee recommendation also provides $222,227,506,000 in advance appropriations for the Veterans Benefits Administration's mandatory accounts for fiscal year 2026. This level may be adjusted based on the results of the administration's mid-session review.

## Compensation and Pensions (Including Transfer of Funds)

| | |
|---|---|
| Appropriations, 2024 | $161,850,524,000 |
| Advance Appropriations, 2025 | 182,310,515,000 |
| Budget estimate, 2025 | 9,820,699,000 |
| Committee recommendation, 2025 | 9,820,699,000 |
| Budget estimate, advance appropriations, 2026 | 204,481,753,000 |
| Committee recommendation, advance appropriations, 2026 | 204,481,753,000 |

## Program Description

Compensation is payable to living veterans who have suffered impairment of earning power from service-connected disabilities. The amount of compensation is based upon the impact of disabilities on a veteran's earning capacity. Death compensation or dependency and indemnity compensation is payable to the surviving spouses and dependents of veterans whose deaths occur while on active duty or result from service-connected disabilities. A clothing allowance may also be provided for veterans with a service-connected disability who use a prosthetic or orthopedic device. In fiscal year 2025, the Department-estimates-it-will-obligate $188,743,668,000 for compensation payments to 6,312,645 veterans, 558,394 survivors, and 1,057 children receiving special benefits.

Pensions are an income security benefit payable to certain wartime veterans who are precluded from gainful employment due to non-service-connected disabilities that render them permanently and totally disabled. The Veterans Education and Benefits Expansion Act of 2001 (Public Law 107–103), restored the automatic presumption of permanent and total non-service-connected disability for purposes of awarding a pension to veterans age 65 and older, subject to the income limitations that apply to all pensioners. Death pensions are payable to certain surviving spouses and children of deceased wartime veterans. The rate payable for both disability and death pensions is determined on the basis of the annual income of the veteran or their survivors. In fiscal year 2025, the Department estimates that the pensions program will provide benefits to 129,417 veterans and 94,884 survivors totaling $3,038,638,000.

The Compensation and Pensions program funds certain burial benefits on behalf of eligible deceased veterans. These benefits provide the purchase and transportation costs for headstones and markers, graveliners, and pre-placed crypts; and provide partial reimbursement for privately purchased outer burial receptacles. In fiscal year 2025, the Department estimates the Compensation and Pensions program will obligate $348,908,000 providing burial benefits. This funding will provide 40,183 burial allowances, 48,005 burial plot allowances, 60,512 service-connected death awards, 401,648 burial flags, 327,779 headstones or markers, 49,715 graveliners or reimbursement for privately purchased outer burial receptacles, 232 caskets and urns for the internment of the remains of veterans without next of kin, and 1,673 cremation urns and commemorative plaques in lieu of furnishing a headstone or marker for an eligible deceased individual not already buried who served in the Armed Forces.

## Committee Recommendation

In fiscal year 2024, the Committee provided an advance appropriation of $182,310,515,000 for fiscal year 2025 for the Compensation and Pensions account.

The Committee recommendation includes an additional $9,820,699,000 for fiscal year 2025 and an advance appropriation of $204,481,753,000 for the Compensation and Pensions account for fiscal year 2026, as requested.

*Oversight of Compensation and Pension [C&P] Exams.* —The Committee is aware that the Department has increasingly relied on outside providers to conduct C&P exams. Additional oversight is necessary to ensure veterans are receiving quality exams and that they are easily accessible. The Department should apply consistent standards for timeliness and quality performance for both contract providers and VHA providers. Additionally, VA should maintain statistics regarding the rate and frequency with which contract providers and VA render negative opinions regarding C&P exams to monitor for trends. At a minimum, such statistics should contain provider name, facility, contractor or Federal employee status, and list of conditions (with corresponding rate of diagnosis) for which the provider performs C&P exams.

Further, the Department should utilize the terms of these contracts to institute more robust oversight mechanisms, including setting up a process by which veterans can lodge complaints with VBA that the agency then can resolve directly with the contractor.

Additionally, the Committee directs VA to report to the Committees on Appropriations of both Houses of Congress on steps it will take to implement the recommendations in the VA Office of Inspector General's report ''The Medical Disability Examination Office Needs to Better Monitor Mileage Requirements for Contract Exams'' dated April 20, 2023.

*Unaccredited Claims Agents.*—The Committee is concerned with the proliferation of unaccredited entities, some of whom are aggressively marketing representational services to veterans using predatory tactics, often resulting in veterans paying exorbitant fees for representation for disability compensation claims. The Committee suggests that the Department consider instituting a system of reporting the conduct of unaccredited attorneys to the appropriate bar associations as another means of policing bad conduct. While the Department lacks the authority to do more than send cease and desist letters to entities engaged in these activities, VA is directed to report to the Committees on Appropriations of both Houses of Congress on the steps VA plans to take on instituting a system of reporting the conduct of unaccredited attorneys, the steps it has undertaken to both inform veterans about unaccredited claims representatives and to educate veterans on the risks of sharing their VA account log in credentials and bank account information with such actors.

*Claims and Appeals Backlogs.*—The Committee is concerned about service delays, processing backlogs, and accuracy rates of decisions. The Department is directed to provide a report to the Committees on Appropriations of both Houses of Congress within 180 days of enactment of this act to identify where in the current claims and appeals processes delays and inaccuracies in decisions are occurring, potential personnel shortfalls, and employee training that could be improved upon to increase accuracy and efficiency in claims decisions. The Department is further directed to include information about the average time for processing claims requested for ''Higher-Level Review'' and what efforts could be undertaken to decrease these timelines.

*Military Sexual Trauma [MST].*— Data from VA screenings show that about one in three women veterans experienced MST at some point during their service. In 2018, an OIG report found that when women veterans file claims for MST, nearly half of them are denied disability benefits. Since then, Congress enacted legislation to require specialized teams at VBA to process MST-related claims. Therefore, the Committee encourages the Secretary to determine whether it is possible to further streamline the claims process for veterans who reported sexual assault that occurred during military service to limit the number of engagements with claims processors that require such veterans to repeat their experience multiple times. Further, the Committee encourages the Secretary to examine, as available, the potential differences between rating outcomes for veterans who reported their sexual assault while in service and those who did not. The Secretary should report its findings to the Committees on Appropriations of both Houses of Congress no later than 120 days after enactment.

## Readjustment Benefits

| | |
|---|---|
| Appropriations, 2024 | $8,827,352,000 |
| Advance Appropriations, 2025 | 13,399,805,000 |
| Budget estimate, 2025 | 2,657,656,000 |
| Committee recommendation, 2025 | 2,657,656,000 |
| Budget estimate, advance appropriations, 2026 | 17,614,235,000 |
| Committee recommendation, advance appropriations, 2026 | 7,614,235,000 |

## Program Description

The Readjustment Benefits appropriation finances the education and training of veterans and servicemembers under chapters 21, 30, 31, 33, 34, 35, 36, 39, 41, 51, 53, 55, and 61 of title 38, United States Code. These benefits include the All-Volunteer Force Educational Assistance Program (Montgomery GI bill) and the Post-9/ 11 Educational Assistance Program, including changes through the Forever GI bill. Basic benefits are funded through appropriations made to the readjustment benefits account and by transfers from the Department of Defense [DoD]. This account also finances vocational rehabilitation, specially adapted housing grants, specially adapted automobile grants for certain disabled veterans, and educational assistance allowances for eligible dependents of those veterans who died from service-connected causes or who have a total permanent service-connected disability, as well as dependents of servicemembers who were captured or missing in action.

## Committee Recommendation

In fiscal year 2024, the Committee provided an advance appropriation of $13,399,805,000 for the Readjustment Benefits account for fiscal year 2025.

The Committee recommendation includes an additional $2,657,656,000 for fiscal year 2025 and an advance appropriation of $17,614,235,000 for Readjustment Benefits for fiscal year 2026, as requested.

*Excess Leave Program [ELP].*—The Committee is aware that VA stopped providing the Military Housing Allowance benefit for a cohort of active duty service men and women attending university under the ELP. The Committee urges the Department to consider utilizing discretionary authority to begin providing this housing benefit to participants in this program.

*Postsecondary Credential among Veterans.*—The Committee directs the Comptroller General to report on the prevalence of high-quality postsecondary credentials among veterans. The report shall include the number of veterans who hold a high-quality postsecondary credential, the number of those veterans who obtained such a credential during their military service, the number of those veterans who obtained such a credential after their military service, an earning comparison between veterans with and without such credentials who are working in similar industries and occupations, the rate of business ownership among veterans with and without such credentials, the average cost and time needed for veterans to obtain such credentials, and recommendations for how to increase veterans' access to such credentials. The Comptroller General shall consider credentials that are: (i) included on a State list described in section 122(d) of the Workforce Innovation and Opportunity Act (29 U.S.C. 3152(d)); (ii) in compliance with the quality assurance requirements for a credentialing program under 10 U.S.C. 2015(c) (i.e., the COOL directory); (iii) approved by the State approving agency according to the Department of Veterans Affairs WEAMS database; (iv) accredited by the Institute for Credentialing Excellence, the National Commission on Certifying Agencies, the American National Standards Institute, or other reputable accrediting bodies; and (v) meet other standards for quality determined by the Departments. The Comptroller General shall provide a report to the Committees on Appropriations of both houses of Congress no later than 1 year after the enactment of this act.

*Veteran Readiness and Employment [VR&E] Case Managers.*—Veterans report experiencing difficulty in contacting a case manager for help related to VR&E programs. The Department must provide access to assistance needed in a timely manner. Therefore, the Department is directed to provide the Committees on Appropriations of both Houses of Congress a report on wait times for a veteran to reach a case manager, and what resources the Department needs to bring wait times within a timeframe the Secretary deems satisfactory. The report should be provided within 180 days of enactment of this act, and discuss the VR&E staff ratio, if the current requirement of 1 FTE to 125 veterans is sufficient, the geographical spread of VR&E counselors and its impact on serving veterans, and constraints on access to VR&E counselors in rural communities.

*Educating Student Veterans on Debt.*—The Committee encourages the Department to share additional detailed information with students and Institutions of Higher Learning [IHL] on student debt, including debt incurred against IHL for overpayments to the student for tuition and fees or

Yellow Ribbon payments, for the purposes of working with such students and institutions to eliminate the debt.

*GI Bill Requirements for Flight Training.*—In 2015 VA issued guidance requiring all GI Bill-approved Institutions of Higher Learning [IHL] to establish a specific number of training hours required for completion of educational objectives for all students. The required number of training hours must be stated as a specific number and may not be presented as a range or a maximum or minimum. The Committee is concerned that this requirement may not allow for necessary flexibility in a competency-based curriculum and could force veterans to retake the course using their VA educational benefit. The Committee directs the Department to consider increasing flexibility so that flight schools affiliated with IHL are not mandated to fail students who complete coursework before or after the specific training hours requirement as well as establishing a standard based on an allotted dollar amount per student rather than an hour-based requirement. The Department shall submit a report to the Committees on Appropriations of both Houses of Congress on the advisability, feasibility, and cost of taking such actions no later than 180 days after enactment of this act. *Biomanufacturing Training.*—The Committee directs the Department to consider engaging with an institution of higher education for the provision of biomanufacturing training to provide targeted, hands-on, industrial biomanufacturing training for veterans and transitioning service members.

*Technology Education Programs.*—The Committee is encouraged by the success of the VET TEC program and encourages the Department to take any steps available to provide veterans with similar technology education programs, particularly in areas of the country where such initiatives have not been previously available.

**Veterans Insurance and Indemnities**

| | |
|---|---|
| Appropriations, 2024 | $133,827,000 |
| Advance Appropriations, 2025 | 135,119,422 |
| Budget estimate, advance appropriations, 2026 | 131,518,000 |
| Committee recommendation, advance appropriations, 2026 | 131,518,000 |

## Program Description

The Veterans Insurance and Indemnities appropriation consists of the former appropriations for military and naval insurance, applicable to World War I veterans; National Service Life Insurance, applicable to certain World War II veterans; servicemen's indemnities, applicable to Korean conflict veterans; and veterans mortgage life insurance to individuals who have received a grant for specially adapted housing.

## Committee Recommendation

In fiscal year 2024, the Committee provided an advance appropriation of $135,119,422 for fiscal year 2025 for the Veterans Insurance and Indemnities account. In addition, the Committee recommendation includes an advance appropriation of $131,518,000 for Veterans Insurance and Indemnities for fiscal year 2026, as requested.

Veterans Housing Benefit Program Fund

| | Program account[1] | Administrative expenses |
|---|---|---|
| Appropriations, 2024 ........ | SSAN | $316,742,419 |
| Budget estimate, 2025 ........ | SSAN | 319,596,460 |
| Committee recommendation, 2025 ........ | SSAN | 319,596,460 |

[1]Such sums as may be necessary [SSAN].

## Program Description

The Veterans Housing Benefit Program Fund provides for all costs associated with VA's direct and guaranteed housing loan programs, with the exception of the Native American Veteran Housing Loan Program.

VA loan guaranties are made to servicemembers, veterans, reservists, and unremarried surviving spouses for the purchase of homes, condominiums, and manufactured homes, and for refinancing loans. VA guarantees part of the total loan, permitting the purchaser to obtain a mortgage with a competitive interest rate, even without a downpayment, if the lender agrees. VA requires a downpayment be made for a manufactured home. With a VA guaranty, the lender is protected against loss up to the amount of the guaranty if the borrower fails to repay the loan.

## Committee Recommendation

The Committee recommends such sums as may be necessary, currently estimated to be $2,089,000,000, for funding subsidy payments, and $319,596,460 for administrative expenses for fiscal year 2025. Bill language limits gross obligations for direct loans for specially adapted housing to $500,000.

*Veterans Assistance Partial Claim Payment [VAPCP] Program.*— The Committee recognizes the hardship that the abrupt end of the VAPCP program had on thousands of veteran homeowners, and urges the Department to provide immediate assistance to veterans who were impacted by increased mortgage payments caused by modification of existing loans. The Committee directs the Department to provide a report to the Committees on Appropriations of both Houses of Congress, no later than 90 days after enactment of this act, with information on efforts undertaken to assist veterans impacted by the abrupt end of the VAPCP program. The report shall include an overview of the implementation of the Veterans Affairs Service Purchasing Program and identify services available and needed to counsel veterans on avoiding or recovering from delinquency and foreclosure.

*Revised Energy Standards.*—Certain Federal agencies have recently adopted revised energy standards for newly constructed and rehabilitated homes insured or guaranteed by their respective Departments. Within 180 days of enactment, the Department is directed to provide the Committee with its own evaluation of how adoption of the revised energy standards would impact costs and veteran homebuyers, including the availability, affordability, and competitiveness of VA home loans.

## Vocational Rehabilitation Loans Program Account

| | Program account | Administrative expenses |
|---|---|---|
| Appropriations, 2024 | $78,337 | $460,698 |
| Budget estimate, 2025 | 64,431 | 493,868 |
| Committee recommendation, 2025 | 64,431 | 493,868 |

**Program Description**

The Vocational Rehabilitation Loans Program account covers the cost of direct loans for vocational rehabilitation of eligible veterans and, in addition, includes administrative expenses necessary to carry out the direct loan program. Loans of up to $1,530 (based on the indexed chapter 31 subsistence allowance rate) are currently available to service-connected disabled veterans enrolled in vocational rehabilitation programs, as provided under 38 U.S.C. chapter 31, when the veteran is temporarily in need of additional assistance. Repayment is made in monthly installments, without interest, through deductions from future payments of compensation, pension, subsistence allowance, educational assistance allowance, or retirement pay. Virtually all loans are repaid in full and most in less than 1 year.

**Committee Recommendation**

The Committee recommends $64,431 for program costs and $493,868 for administrative expenses for the Vocational Rehabilitation Loans Program account. The administrative expenses may be paid to the General Operating Expenses, Veterans Benefits Administration account. Bill language is included limiting program direct loans to $1,563,660. It is estimated VA will make 1,000 loans in fiscal year 2025, with an average amount of $1,303.

**Native American Veteran Housing Loan Program Account**

| | |
|---|---|
| Appropriations, 2024 | $2,718,546 |
| Budget estimate, 2025. | 5,845,241 |
| Committee recommendation. | 5,845,241 |

**Program Description**

The Native American Veteran Housing Loan Program is authorized by 38 U.S.C. chapter 37, section 3761 to provide direct loans to Native American veterans living on trust lands. The loans are available to purchase, construct, or improve homes to be occupied as veteran residences or to refinance a loan previously made under this program in order to lower the interest rate.

Veterans pay a funding fee of 1.25 percent of the loan amount, although veterans with a service-connected disability are exempt from paying the fee. Before a direct loan can be made, the veteran's Tribal organization must sign a memorandum of understanding with VA regarding the terms and conditions of the loan. The Native American Veteran Housing Loan Program began as a pilot program in 1993 and was made permanent by the Veterans' Housing Opportunity and Benefits Improvement Act of 2006 (Public Law 109–233).

## Committee Recommendation

The Committee recommends $5,845,241 for administrative expenses associated with this program, as requested.

*Native American Veteran Housing Loan Program*.—The Committee recognizes that there is a tremendous need for safe and affordable housing in American Indian, Alaska Native, and Native Hawaiian communities. Native American Community Development Financial Institutions [NCDFIs] have deep ties to the local communities they serve, and are uniquely equipped to reach potential homebuyers. The Secretary should consider allowing Tribally Designated Housing Authorities and NCDFIs to act as intermediary lenders in Tribal communities and for the program to collaborate with these entities to provide greater homebuyer education. The Secretary is directed to conduct meaningful and periodic outreach to Tribes, Alaska Native entities, and Native Hawaiian entities to increase geographic diversity of awards and increase awareness and enrollment among Native American veterans.

## General Operating Expenses, Veterans Benefits Administration

| | |
|---|---|
| Appropriations, 2024 | $3,899,000,000 |
| Budget estimate, 2025 | 4,035,000,000 |
| Committee recommendation | 4,035,000,000 |

## Program Description

The General Operating Expenses [GOE], VBA account provides funding for VBA to administer entitlement programs such as service-connected disability compensation, education benefits, and vocational rehabilitation services.

## Committee Recommendation

The Committee recommends $4,035,000,000 for the GOE, VBA account, which is $136,000,000 above the fiscal year 2024 enacted level and equal to the budget request. The Committee included bill language to make available through September 30, 2026, up to 10 percent of the GOE, VBA account.

In addition, VA estimates to use $1,401,000,000 from the Toxic Exposures Fund to support GOE, VBA-related activities in fiscal year 2025.

*Availability of GI Bill Benefits*.—The Committee notes that VA policy determinations have restricted the ability of veterans to access their earned benefits, including GI Bill Benefits. In particular, the Committee is aware of the VA Policy Advisory, dated September 2, 2022 and entitled ''State-Legalized Cannabis Training and GI Bill Benefits and Related Addendum,'' which changed VA policy to no longer allow the use of GI Bill Benefits for courses of study determined to ''have the objective to prepare someone to participate in the cultivation, sale, or distribution of marijuana.'' The Advisory came despite the offering of such courses in States in which either the recreational or medicinal use of marijuana is legally permissible. The Committee directs VA to submit a report to the Committees on Appropriations of both Houses of Congress no later than 90 days after enactment of this act regarding the number of veterans who this effected.

*VA Apprenticeship and On-the-Job-Training [OJT]*.—The Committee notes that the fiscal year 2023 Joint Explanatory Statement accompanying the Consolidated Appropriations Act, 2023 directed VA to ''take additional actions to promote awareness and increased utilization of apprenticeships and OJT programs'' available to veterans, including, in coordination with the Department of Labor, the establishment of a user-friendly website on which veterans could find information about apprenticeship programs approved under the GI Bill. Additionally, VA was encouraged to continue coordination with State Approving Agencies [SAA] on programs which may incentivize increased participation by employers. Within 180 days of enactment of this act, the Department is directed to provide an update to the Committees on

Appropriations of both Houses of Congress regarding its efforts to develop the website and actions it has taken with SAAs for the purpose of increasing awareness and utilization of apprenticeships and OJT.

*Outreach to Upgrade-Eligible Veterans.*—The Committee supports efforts to ensure that veterans can access the benefits to which they are entitled following their service to our Nation. In particular, the Committee recognizes the importance of ensuring that veterans who were given a less than honorable discharge from the military due to behaviors resulting from traumas such as PTSD, TBI, or MST are aware of the opportunity to upgrade their service discharge status via petitioning of a Discharge Review Board, which each military service operates. The Department is directed to provide a report to the Committees on Appropriations of both Houses of Congress within 180 days regarding efforts already underway by the Department to conduct outreach towards veterans who may be eligible for a discharge upgrade to ensure they are aware of the review board petition process. Such report shall also include recommendations regarding opportunities for additional and more effective outreach to veterans, including, but not limited to, efforts that could be effectuated via the VA–DOD Joint Executive Committee to enhance coordination between VA and DOD to ensure both veterans and those soon to leave the service are aware of the Discharge Review Board process and the requirement of liberal consideration under Public Law 114–328, Section 535.

Further, within 1 year of enactment, VA shall submit a report to the Committees on Appropriations of both Houses of Congress detailing the Department's outreach efforts towards LGBTQ∂ veterans, what VA has done to partner with State Veterans Administrations and VSOs and what resources VA makes available to State Veterans Administrations on LGBTQ∂ outreach with regards to the upgraded discharge process.

*Support to County, Tribal, and Equivalent Governmental Veterans Service Officers.*—The Committee recognizes the crucial role of County, Tribal and equivalent governmental veterans service officers in ensuring veterans receive the benefits and care they deserve. Accordingly, the Committee looks forward to receiving the report on the progress made to develop a systemic method for facilitating collaboration between County, State, and Tribal veterans service officers, VHA facilities, and VBA offices in their respective geographic areas requested in S. Rept. 118–43.

*VetSuccess on Campus Program.*—The Committee recognizes the importance of the VetSuccess on Campus program and its role in ensuring success in post-secondary education for veterans, service members, and qualified dependents. The Committee directs the agency to seek opportunities for program expansion, including locations that serve multiple institutions and in States that currently do not benefit from the program.

## Veterans Health Administration

| | |
|---|---|
| Appropriations, 2024 | $121,011,303,000 |
| Advance appropriations, 2025 | 112,582,000,000 |
| Budget estimate, 2025 | |
| Committee recommendation, 2025. | 3,000,000 |
| Budget estimate, advance appropriations, 2026 | 131,439,000,000 |
| Committee recommendation, advance appropriations, 2026 | 131,439,000,000 |

## Administration Overview

The Veterans Health Administration [VHA] is home to the United States' largest integrated healthcare system.

The Department of Veterans Affairs Medical Care Collections Fund [MCCF] was established by the Balanced Budget Act of 1997 (Public Law 105–33). In fiscal year 2004, the Consolidated Appropriations Act, 2004 (Public Law 108–199) allowed the Department to deposit first-party and pharmaceutical co-payments, third-party insurance payments and enhanced-use collections, long-term care co-payments, Compensated Work Therapy Program collections, Compensation and Pension Living Expenses Program collections, and Parking Program fees into the MCCF.

The Parking Program provides funds for the construction, alteration, and acquisition (by purchase or lease) of parking garages at VA medical facilities authorized by 38 U.S.C. 8109. The Secretary is required under certain circumstances to establish and collect fees for the use of such garages and parking facilities. Receipts from the parking fees are to be deposited into the MCCF and are used for medical services activities.

## Committee Recommendation

In fiscal year 2024, the Committee provided $112,582,000,000 in advance appropriations for VA's medical care accounts for fiscal year 2025. This included $71,000,000,000 for Medical Services, $20,382,000,000 for Medical Community Care, $11,800,000,000 for Medical Support and Compliance, and $9,400,000,000 for Medical Facilities. VA did not request a "second bite" for medical care accounts in fiscal year 2025.

For fiscal year 2025, the Committee provides a total of $112,585,000,000 for VA medical care. Additionally, the Committee recommendation includes $878,000,000 for Medical and Prosthetic Research. Medical Care Collections are estimated to be $4,632,000,000 in fiscal year 2025. The Committee recommendation also provides $131,439,000,000 in advance appropriations for VA's medical care accounts for fiscal year 2026, consistent with the budget request.

The Fiscal Responsibility Act (Public Law 118–5) directly appropriated $24,455,000,000 for the Toxic Exposures Fund, of which $21,454,542,000 is estimated to be for medical care, for a total of $134,039,542,000 in medical care funding in fiscal year 2025, $3,000,000 over the budget request.

The Committee recognizes the invaluable role of the Department of Veterans Affairs, the Nation's largest integrated healthcare system, in serving the unique needs of veterans. The Committee recognizes that a strong and fully resourced VHA is necessary to effectively serve our Nation's veterans.

*Childcare Infrastructure.*—The Committee provides $3,000,000 for costs associated with infrastructure supporting the Department's childcare initiative and requires a spend plan to be submitted to the Committees on Appropriations of both Houses of Congress within 90 days of enactment of this act for these funds.

*Child Care Assistance.*—The Committee continues to support the Department's planned efforts to expand child care assistance to veterans receiving healthcare at all VA medical facilities, and is disappointed by the lack of progress made thus far. The Department is reminded of the statutory deadline of January 5, 2026, as required by the Johnny Isakson and David P. Roe, M.D. Veterans Health Care and Benefits Improvement Act of 2020 (Public Law 116–315), and thus is directed to expeditiously publish regulations and to provide quarterly updates to the Committees on Appropriations of both Houses of Congress on the implementation of this program, including what VA is doing to inform veterans of these services.

*Vet Centers.*—The Department is directed to submit to the Committees on Appropriations of both Houses of Congress, within 90 days of enactment of this act, a report detailing its progress in meeting the statutory deadlines under the Vet Center Improvement Act (Public Law 117–263), and if additional funds are required to support full implementation.

*Records Management.*—The Committee directs VA to assess options for modernizing medical records storage at the VA Records Management Center, and urges the Department to make available options to send copies of records other than on compact discs.

## Medical Services (Including Transfer of Funds)

| | |
|---|---|
| Appropriations, 2024 | $70,969,795,000 |
| Advance appropriations, 2025 | 71,000,000,000 |
| Budget estimate, advance appropriations, 2026 | 75,039,000,000 |
| Committee recommendation, advance appropriations, 2026 | 75,039,000,000 |

## Program Description

The Medical Services account provides for medical services for eligible enrolled veterans and other beneficiaries in VA healthcare facilities, including VA medical centers and VA outpatient clinics.

## Committee Recommendation

In fiscal year 2024, the Committee provided an advance appropriation of $71,000,000,000 for fiscal year 2025 for the Medical Services account. In addition, VA estimates to use $11,683,896,000 from the Toxic Exposures Fund to support medical services in fiscal year 2025. VA also has the authority to retain co-payments and third-party collections.

The Committee recommendation also includes an advance appropriation of $75,039,000,000 for Medical Services for fiscal year 2026, equal to the budget request.

## Preventing Veteran Homelessness

The Committee remains strongly supportive of VA's homelessness prevention programs. As such the recommendation includes $3,210,276,000 to support these programs, $96,746,000 more than fiscal year 2024 levels, and equal to the budget request. This total includes $659,049,000 for the Supportive Services for Veterans Families [SSVF] Program; $661,535,000 for Housing and Urban Development-Veterans Affairs Supportive Housing [HUD–VASH] Program case management, $319,788,000 for the Grant and Per Diem [GPD] Program; and $152,793,000 for the Veterans Justice Outreach [VJO] Program and Legal Services for Veterans grants. The Committee recognizes the Department's efforts to reduce the number of veterans experiencing homelessness, and the goal of putting 41,000 veterans experiencing homelessness into long term housing in calendar year 2024.

*HUD–VASH Vouchers and Case Managers.*—The Committee recognizes HUD–VASH as an important tool to house homeless veterans. However, the overall eligibility and referral processes can vary due to local coordination and resource availability. The Committee therefore directs the Department, in coordination with the Department of Housing and Urban Development, to provide a report to the Committees on Appropriations of both Houses of Congress no later than 180 days after enactment of this act detailing how the Departments are developing a coordinated entry system, including allocations of yearly vouchers, and how the Departments plan to reduce the number of unused vouchers. The Committee further directs VA to provide quarterly updates on the number of HUD–VASH case managers directly employed by the Department, the number of contracted positions, and the average caseload of each case manager by VISN.

*Supportive Services for Veteran Families Program [SSVF].*—The Committee urges VA to take steps to ensure that the SSVF program continues to meet the needs of veterans, including the impact of rising costs for household goods. The Department should consider increasing the cap on Tenant Incentives and to consistently cover costs associated with pet deposits, rent, utility deposits, and car repairs.

*Connected Care Program.*—The Connected Care Program provides valuable resources, such as smartphones and tablets that help to ensure that homeless veterans have access to telehealth and other online services. Based on a 2022 Inspector General report, however, the Committee is concerned that the program is not effectively utilizing its resources. Therefore, the Committee encourages the Secretary to provide a report to Committees on Appropriations of both Houses of Congress no later than 120 days after enactment of this act identifying the steps taken by the Department to improve and expand the program to better serve homeless veterans.

## Telehealth Services

The Committee recommendation includes $6,398,020,000, $1,232,684,000 over fiscal year 2024 levels, and equal to the budget request, to sustain and increase telehealth capacity, including in rural and highly rural areas. VA and DoD have long been leaders in the field of emerging technology and medicine. For VA, telehealth creates a bridge between rural and urban centers, allowing the Department to expand access to care in areas where services are limited. Telehealth also allows care to be provided more effectively and efficiently for veterans closer to home and through direct in-home access to providers. While VA continues to lead the healthcare industry in the expansion of in-home telehealth and remote patient monitoring services, these services are often limited by the lack of broadband service in remote and rural areas.

*Innovative Telehealth Services.*—The Committee understands that telehealth has been a successful tool in helping veterans access healthcare through VA. The Committee encourages the Department to continue increasing the reach of VA providers to veterans in rural areas through the use of telehealth, especially for primary care and preventative care services. Within 90 days of enactment of this act, the Department shall provide a spend plan to the Committees on Appropriations of both Houses of Congress for the funding provided in support of telehealth initiatives, including an overview of innovative efforts to reach veterans in rural areas with telehealth.

*Ketogenic Nutrition Programs.*—The Committee understands the Department is using telehealth services as a means to provide individualized ketogenic nutrition programs and medical support to reduce the incidence of diabetes in VA patients. The Committee encourages the Department to consider expanding such efforts to address the healthcare needs of Veterans living with Type 2 diabetes.

## Mental Health/Preventing Veterans Suicide

The Committee provides $16,380,187,000 for mental health programs, equal to the budget request, including $3,250,267,000 for suicide prevention outreach and treatment programs, of which $582,554,000 is for suicide prevention outreach.

*Improving Depression Treatment with Precision Medicine.*—The Committee recognizes that depression is one of the most common conditions associated with military service and combat exposure. The Department is directed to provide a report to the Committees on Appropriations of both Houses of Congress on ongoing or upcoming partnerships providing pharmacogenomic tests to inform treatments for clinical depression.

*Lethal Means Safety and Suicide Prevention.*—The Committee is deeply concerned by recent increases in suicide rates and the prevalence of firearm-related suicides across the veteran community— particularly striking among women veterans. To combat these trends, the Committee supports VA's efforts related to lethal means safety and safe storage programming. The Committee directs the Department to expand these efforts as part of the National Strategy for Preventing Veteran Suicide, to enable greater distribution of firearms safe storage devices. Further, the Committee directs VA to ensure that access to firearms storage resources for veterans is included in training for employees of the Veterans Crisis Line. The Secretary is directed to submit an annual report to the Committees on Appropriations of both Houses of Congress, disclosing the number of veterans making requests for safe storage devices, the number of safe storage devices distributed, the funding needs for the continued operation of the program, and an assessment of the cost of educating all veterans who receive care at VA facilities about the availability of safe storage devices as a means of suicide prevention and harm reduction.

*Post-Traumatic Stress Disorder [PTSD] Best Practices.*—The Committee recognizes that VA has significant knowledge and understanding as it relates to development of screening, educational materials, and treatment of PTSD. The Committee encourages the Secretary to share best practices around PTSD screenings, education, and treatment with other Federal agencies as appropriate. *Predictive Modeling and Analytics for Veterans Suicide Prevention.*—Suicide is a complex issue that requires a comprehensive and innovative solution. VA has already started to tackle this by creating Recovery Engagement and Coordination for Health—Veterans Enhanced Treatment [REACH VET], an identification and intervention program that uses predictive analytics to identify at-risk veterans. The Committee is aware of additional predictive data analytics and machine learning tools that may help at-risk veterans before a crisis arises. The Committee encourages VA to use predictive data analytics and machine learning more broadly across the system to identify veterans with suicidal ideations and better deliver treatment.

*Psychedelic-Assisted Therapy.*—The Committee understands that VA and other relevant Federal agencies are undertaking research to evaluate the efficacy of psychedelic-assisted therapies in treating PTSD, major depressive disorder, and other conditions. The Committee directs the Secretary to submit a report no later than 180 days after enactment of this act to the Committees on Appropriations of both Houses of Congress, on current research and activities related to these therapies.

Further, the Department is directed to initiate a longitudinal study of veterans participating in such therapies and track outcomes over a period of 5 years. The study should include: the number of individuals receiving psychedelic-assisted therapies treatment in the last year, the average number of months such individuals served on active duty if available, the distribution of disability ratings of such individuals, the gender distribution of individuals receiving treatment, the number of individuals receiving psychedelic-assisted therapies treatment who suspended participation in such treatment, the average number of treatment sessions and dosages each individual received, the percentage of individuals who experienced a clinically significant reduction in symptoms, and the number of such individuals who experienced a recurrence of their diagnosis after previously receiving this treatment. The Department is directed to report annually on the progress of the study and provide a report to the Committees on Appropriations of both Houses of Congress on outcomes, at the conclusion of the study.

*Veterans Crisis Line [VCL].*—The Committee notes that engagement with VCL has increased, and is concerned that some veterans seeking crisis intervention services may not have received the care or connection to local providers that they needed. It is critical that the VCL has the oversight and training necessary to support a substantial rise in call volume. Therefore, the Committee directs the Secretary to report within 90 days of enactment of this act to the Committees on Appropriations of both Houses of Congress on the implementation of the VA Office of the Inspector General recommendations included in the September 13, 2023, VA OIG report, 22–00507–211.

*Natural Language Speech.*—The Committee encourages the Secretary to implement ''Natural Language Speech'' selection capability for contact center solutions across VA Health Contact Centers including the Veteran's Crisis Line and the Community Care Enterprise Contact Center, as practicable.

*Farming and Recovery Mental Health Services.*—The Committee recognizes that the Department's Farming and Recovery Mental Health Services program has been effective in training veterans in agricultural vocations while also addressing behavioral and mental health needs. However, only six of the original 10 sites of the pilot program are still active today. The Committee therefore encourages the Department to reevaluate the program and to consider expansion.

*Veteran Provider De-escalation Training.*—The Committee recognizes that training with virtual reality or augmented reality technologies can improve provider awareness on how to de-escalate potentially volatile situations associated with caring for veterans with post-traumatic stress disorder, traumatic brain injury and other trauma. The Committee encourages VA to consider using these novel technologies to provide de-escalation training to appropriate VA providers.

## Opioid Safety Initiatives and Substance Use Disorder Care

To continue to build upon opioid reduction efforts and safety initiatives, the Committee recommendation includes $713,049,000 for Opioid Prevention and Treatment programs at VA, equal to the budget request. This includes $458,562,000 for treatment programs and $254,487,000 to continue implementation of the Jason Simcakoski Memorial and Promise Act, as part of the Comprehensive Addiction and Recovery Act of 2016 (Public Law 114–198).

*Non-Addictive Opioid Alternatives.*—The Committee remains concerned about the continued use of opioids in the veteran population. The Committee directs the Department to provide a report to the Committees on Appropriations of both Houses of Congress no later than 60 days after enactment of this act on veterans' access to non-opioid pain relief medications. The report should include steps the Department will take to ensure that coverage, cost-sharing, and access are not an impediment to veterans receiving these medications, what actions need to be taken to ensure enrolled veterans have access to prescriptions for nonaddictive opioid alternatives for both acute and chronic pain, and how the Department will support continued and expanded education to providers on non-addictive opioid alternatives.

*Training and Medication Access.*—The Committee is concerned that veterans experiencing substance use disorder are not able to access treatments, such as buprenorphine. The Committee directs VA to expand training and education for providers on the use of FDA-approved medication assisted treatments for substance use disorder, and to coordinate with provider organizations to ensure that veterans are able to access the medications for substance use disorder in inpatient and emergency settings.

*Pain Management Decision Support Tool.*—The Committee recognizes the balance between supporting access to care for veterans with pain and the veteran's risk of developing dependency for prescription pain medication. The

Committee supports efforts to establish more precise and effective prescribing guidelines to ensure veterans receive appropriate care, and encourages VA to consider deploying a clinical decision support tool to evaluate and interpret the acute and chronic pain management needs for eligible veterans. The Department is directed to provide a report within 180 days on the feasibility, advisability and cost estimates for implementing such a tool.

*Osteopathic Manipulative Treatment.*—The Committee notes the report language in the fiscal year 2023 Joint Explanatory Statement requiring the Department to track the use of osteopathic manipulative treatment to treat back and other pain. The Committee continues to encourage VA to expand the use of alternative treatments for pain management, such as acupuncture, in its delivery of healthcare services.

*Pharmacy Benefits Management Services.*—The Committee is aware of new pain management drugs that are less addictive than opioids. VA is encouraged to provide veterans with as many options for pain management besides opioid analgesics therapeutics, including new FDA approved drugs that are proven to be safe and effective.

*Opiate-Free Pain Therapy.*—The Secretary is encouraged to consider implementing a research project evaluating the efficacy of thermal, shortwave diathermy on patients with chronic pain that improves blood flow and provides non-opioid pain relief. The Department is directed to provide a report to the Committees on Appropriations of both Houses of Congress on the feasibility, advisability, and cost estimates of operating such a program.

## Long–Term Care

The Committee provides $17,036,560,000 for institutional and non-institutional care, $4,095,776,000 over fiscal year 2024 and equal to the budget request. The Committee is aware of the aging veteran population and supports long-term care that focuses on facilitating veteran independence, enhancing quality of life, and supporting the family members of veterans. As such, the Committee supports the Department's efforts to broaden veterans' options regarding non-institutional long-term care support and services, and to accommodate veterans' preferences in how and where they receive care and other services.

*Nursing Home, Long-Term, and Hospice Care.*—The Committee understands the growing need for long-term services and supports for veterans and urges the Department to plan accordingly. The Committee applauds the steps taken by VA and Congress to address this gap. The Department is urged to consider new efforts within the Home Maker and Home Health Aid program to address the current workforce shortage. The Committee looks forward to receiving the report directed on an alternate payment structure for providers outlined in S. Rept. 118–43.

*Veterans Access to Long-Term Care.*—The Committee understands the importance of recruiting and retaining vital partners who provide long term care services to veterans. Therefore, the Department should consider working with the Department of Labor to seek exemptions for Veterans Affairs Health Benefits Providers from its authority, to the extent allowable in law.

Community Living Center Beds.—Within 90 days of enactment, the Department is directed to report to the Committees on Appropriations of both Houses of Congress on the feasibility, advisability and costs of increasing the number of operating Community Living Center beds in order to adequately meet the needs of veterans discharged from inpatient care who no longer require acute inpatient care but would benefit from short-term post-discharge stays at a nursing facility.

## Rural Healthcare

Veterans residing in rural and remote areas face unique barriers to receiving high-quality mental health, primary healthcare, and specialty care services. While enhanced community care programs offer veterans increased flexibility to obtain care close to home, there are often gaps in services in rural and remote communities, even among private providers. The Office of Rural Health [ORH] and its Rural Health Initiative has played a critical role in assisting VA in its efforts to increase access to care. Therefore the Committee recommendation includes $342,455,000 for ORH and the Rural Health Initiative, which is $5,000,000 above the budget request.

*Veterans Transportation Service [VTS].*—The Committee understands that transportation to appointments can be a significant barrier to accessing healthcare for veterans, especially in rural areas. The Department is encouraged to require all VA Medical Centers to continue to pursue options to expand partnerships with State and local transportation services, eligible non-profits, and veterans service organizations under the VTS. The Department is directed to submit a report to the Committees on Appropriations of both Houses of Congress no later than 270 days after the enactment of this act detailing the VTS services offered by each VISN, whether existing VTS services are adequate to meet the needs of the veterans served, and opportunities for expanding partnerships with State and local transportation providers, the requirements set by the Department for volunteer drivers to attain and maintain their volunteer driver status, how long on average it takes for a volunteer to be approved after they submit initial paperwork, an assessment of the time it takes for the Department to complete its requirements for new drivers and vehicles such as gas cards and identification production, the number of vehicles out of service for at least 90 days due to repairs needed, the number of hours total driven by volunteers to get veterans to and from medical appointments, and the estimated cost to the Department if those hours needed to be covered by a Department employee.

*Highly Rural Transportation Grants Program.*—The Committee recognizes the need to address the care of veterans living in rural communities across the United States in order to prevent long term disparities in health outcomes. The Committee urges the Department to work with Congress to expand and improve transportation access to and from facilities that serve rural veterans by expanding the definition of ''Highly Rural'' and increasing funds supporting the Highly Rural Transportation Grants Program accordingly.

*Transportation Service Technologies.*—New technologies are being developed and becoming increasingly available in both rural and urban locations throughout the United States which could improve the overall efficiency and coverage of the Veterans Transportation Program. The Committee directs VA to submit a report no later than 180 days after the enactment of this act with an analysis of available data on the impact on homeless veterans from ending the expanded use of the ridesharing program at VA that took place during the COVID–19 pandemic, an estimate of the cost to reinstate the expanded use of the ridesharing program and an identification of any barriers associated with doing so, and an analysis of VA's existing statutory authority for provisions that would enable the agency to restart this program absent legislation from Congress.

*Special Modes of Transportation.*—The Committee directs VA to submit to the Committees on Appropriations of both Houses of Congress an annual report beginning 180 days after date of enactment of this act and annually thereafter for 3 years listing the number of special mode transports conducted within each VISN delineated by those from or to a VA facility, the type of transportation provided, and whether the Department considered the transport emergent.

*Protecting Veterans' Access to Ground and Air Ambulance Services.*—The Committee has concerns that the final rule that reduces reimbursement rates for ground and air ambulance services by linking them to Medicare rates may negatively impact the ability of veterans to access emergency medical services, particularly in rural and underserved areas. Therefore, the bill includes language delaying implementation of the final rule through the end of the fiscal year. The Department should consider further delay of the rule if there will be a negative impact on veterans' healthcare access.

*Outreach to Highly Rural Veterans.*—The Department is directed to conduct outreach to highly rural veterans who lack direct road access to inform them of their eligibility for benefits. Further, the Committee urges the VA Innovation Ecosystem to consider how to expand VA healthcare system's infrastructural reach to enrolled veterans residing in geographically underserved areas, including rural and highly rural veterans.

## Caregivers

The Committee recommendation includes $2,913,000,000 for VA's Caregivers Program, which is $490,590,000 over fiscal year 2024 levels and is equal to the budget request.

The Caregivers Program was enhanced as part of the John S. McCain III, Daniel K. Akaka, and Samuel R. Johnson VA Main-taining Internal Systems and Strengthening Integrated Outside Networks [MISSION] Act of 2018 (Public Law 115–182).

*Program of Comprehensive Assistance for Family Caregivers [PCAFC].*—The Committee commends the Department's improved implementation and expansion of eligibility under the Family Caregiver Program. It is possible that further expansion of the program is warranted, however, as the eligibility testing criteria may not reflect veterans true self-sufficiency. Thus, the Committee requests that VA consider expanding eligibility to services through the PCAFC by modifying such performance testing standards. Additionally, within 30 days of any changes to the program, the Committee directs VA to provide an estimate of increases or decreasesto the number of veterans and caregivers in the program based on those changes.

## Women Veterans Healthcare

The women veteran population is the fastest growing demographic within VA and is anticipated to grow from 800,000 enrolled in 2020 to over 1.2 million by 2030. As a result of the PACT Act, 27 percent of all new enrollees are expected to be women. VA must enhance its services and access to gender-specific care, including primary care providers, gynecologists, maternity care, mammography services, and mental health providers, including care related to experiences of military sexual trauma. Enhancing capacity to support pregnant and postpartum women veterans must also be a priority, which should include expansion of maternity care and lactation support for women veterans.

Toward this end, the Committee recommendation includes $1,323,444,000 for fiscal year 2025, equal to the budget request, to support gender-specific healthcare services, as well as the program office and initiatives. VA is encouraged to use this funding to continue to expand this care so that all women veterans, including those in rural areas, have the access to care they deserve.

*Menopause and Mid-life Women's Health.*—The Committee directs the Department to submit a report to the Committees on Appropriations of both Houses of Congress no later than 120 days after enactment of this act on the care available to women veterans for early detection, diagnosis, and treatment of perimenopause and menopausal symptoms. The report should include any efforts to advance care coordination and expand access to quality primary care and specialty care services with referrals or at VA facilities.

*Mobile Mammography.*—Within 180 days of enactment, the Committee directs VA to perform an analysis of the feasibility and advisability of establishing a mobile mammography program in conjunction with the Indian Health Service and the Department of Defense and report to the Committees on Appropriations of both Houses of Congress on its findings.

*Women Veterans Mental Health Care.*—The number of women veterans accessing VA mental healthcare has nearly doubled over the past decade, and women are now the fastest growing cohort within the veteran community. Based on the upward trend of women in all service branches, the ability for women to serve in combat roles, and the increased complexity of mental health conditions of returning troops, the cost associated with their care will grow accordingly. Therefore, the Committee encourages the Department to continue its support of the National Center for Post-traumatic Stress Disorder Women's Health Sciences Division and the Women's Health Innovation and Staffing

Enhancement Initiative, ensuring adequate resources are forecasted and budgeted for women veterans' mental healthcare.

## Clinical Workforce

The Committee appreciates the Department's efforts to address long-standing challenges in recruiting and retaining physicians, physician assistants, nurses, mental health providers, other healthcare professionals, and related support staff. The Committee reminds the Department of the annual requirement for a report on workforce issues outlined in the Joint Explanatory Statement accompanying Public Law 116–260.

*VHA Staffing.*—The Committee is concerned about existing employee vacancies within VHA and the potential impact on veterans of the elimination of 10,000 full-time staff from the agency. Beginning 90 days after enactment of this act, the Department is directed to provide quarterly updates to the Committees on Appropriations of both Houses of Congress on staffing goals and current staffing levels by VISN, including the number of positions filled or removed in each of these categories: physicians, dentists, registered nurses, licensed practical nurses, nurse assistants, non-physician providers, medical support assistants, and HUD–VASH case managers, and the number of contracted positions in each of these categories nationwide.

*Improving Staffing and Accountability in VHA.*—The Committee notes the recent GAO and VA Office of Inspector General reports highlighting mismanagement in the background investigation process when hiring VA employees, with one in eight employees hired by VHA either not receiving a background check or not receiving one in a timely manner. Therefore, the Committee directs VHA to submit a report to the Committees on Appropriations of both Houses of Congress no later than 180 days after enactment of this act on efforts undertaken to improve the hiring and onboarding process, including background checks, the number of onboarded VHA employees who have not yet completed the background check process, and their efforts to complete the recommendations in the GAO and OIG reports on this topic.

## Other Items of Interest

*VA Intimate Partner Violence Assistance Program [IPVAP].*—The Committee was pleased that the Department included budget details for the IPVAP in the fiscal year 2025 budget submission. The Department is provided $31,373,000 to resource this program in fiscal year 2025. Moreover, the Committee expects that the Department will expand the program as it learns more about the needs of veterans who have suffered intimate partner violence, and directs the Department to report to the Committees on Appropriations of both Houses of Congress on its plans to expand the IPVAP within 180 days of enactment.

*Accessibility of Rape Kits.*—The Committee urges the Department to update all relevant policies and directives to require that VHA emergency departments and urgent care clinics maintain a consistent supply of rape kits, to annually audit the validity of those kits, and to preserve any evidence collected as part of a patient's initial evaluation and stabilization following a sexual assault. The Department is directed to report to the Committees on Appropriations of both Houses of Congress within 180 days regarding any barriers or limitations of the Department in doing so.

*Access to Sexual Assault Forensic/Nurse Examiners [SAFEs/ SANEs].*—The Committee is concerned about the inconsistent availability of treatment by SAFEs/SANEs across VHA emergency departments and urgent care clinics. This lack of reliability can pose an impediment to patients in need of timely medical care and evidence collection following a sexual assault. The Committee directs VHA, not later than 180 days after the enactment of this act, to maintain information on publicly available websites as to where in VA patients may obtain SAFE/SANE services. Such website shall include the facility name, address, and contact information for all VHA facilities where SAFE/SANE services are provided. Within 90 days of enactment, the Committee also directs VA to report to Congress on steps it will take to ensure VA facilities have plans to provide sexual assault forensic examinations to veterans who come to VA facilities after a sexual assault, including those facilities lacking SAFE/SANE resources.

*Colorectal cancer.*—The Committee urges VA to review its policy related to colorectal cancer screening and take action to align it with guideline recommendations from the United States Preventive Services Task Force, National Comprehensive Care Network, and the American Cancer Society. Such action will ensure that veterans have the same degree of access to the most appropriate screening tests as active-duty service members and the general population. Furthermore, the Secretary is directed to provide a report no later than 6 months after enactment of this act to the Committees on Appropriations of both Houses of Congress to receive an update on the alignment of the colorectal screening options available to Veterans.

*Lung Health.*—The Department is encouraged to review and update its approach to lung cancer screening to optimize veterans access, including updating the Department's lung cancer screening guidelines. As part of this process, the Committee suggests the Department seek information and recommendations regarding the latest lung cancer research, data, and clinical best practices. The Department shall provide a report to the Committees on Appropriations of both Houses of Congress within 120 days of enactment of this act regarding the steps it is taking to optimize lung cancer screening for veterans. The Department is further encouraged to consider incorporating non-invasive technologies to screen veterans for respiratory diseases and assess lung function, to limit the need for invasive procedures.

*POPCaP Program.*—The Committee supports the work of the Precision Oncology Program for Cancer of the Prostate [POPCaP] program in providing veterans with access to clinical trials and individually customized treatments based on the cancer's genetic profile and encourages expansion to multiple POPCaP sites at VAMCs across the country.

*Veteran Family Resource Program.*—The Department is directed to provide a report to the Committees on Appropriations of both Houses of Congress, no later than 90 days after enactment of this act, on the status on the Veteran Family Resource Program rollout, including a summary of engagements to date, locations and next steps on rolling the program out to additional VISNs in 2025, and authorities and funding needed to ensure a successful rollout. As part of the Family Resource Program, the Department is encouraged to include information about resources that reduce rates of child abuse and neglect among families with veterans.

*Food Insecurity Training.*—The Department is directed to provide to the Committees on Appropriations of both Houses of Con gress a plan to train veteran-facing staff and direct care providers how to understand the signs of food insecurity so they are able to direct veterans experiencing food insecurity to services available across the Federal Government.

*Adaptive sports.*—The Committee provides $30,542,000 for National Veterans Sports Programs, including $16,000,000 for adaptive sports programs. The Committee recognizes the significant improvements that adaptive sports and recreational therapy have on veterans' mental and physical health, and that veterans have expressed their desire for these activities to be included in the services VA offers. The Committee also recognizes that adaptive sports and recreational therapy can provide a low-cost alternative to other healthcare services that produce similar health outcomes.

*Obstructive Sleep Apnea.*—The Department should consider the advisability and feasibility of establishing a treatment program for veterans suffering from obstructive sleep apnea using an innovative oral appliance that expands the airway.

*Center for Native Hawaiian, Pacific Islander and United States-Affiliated Pacific Islander [NHPIUSAPI] Veterans Health.*—The Committee continues to support VA's Center for NHPIUSAPIs to advance efforts to address the unique healthcare needs of this community. The Center is provided funding consistent with the budget request. The Committee also encourages VA to partner with universities in the Pacific region focusing on issues unique to this community.

Compacts of Free Association Related to Care for Veterans in the Freely Associated States.—To ensure progress in implementation of provisions in Public Law 118–42, 209(a), the Department is directed to provide a report to the Committees on Appropriations of both Houses of Congress, not later than 90 days after enactment of this act, detailing steps taken by the Department thus far, any barriers to implementation, and a cost estimate for the program.

*Produce Prescription Program.*—The Committee recognizes that VHA has the potential to play an important role in alleviating diet-related disease through Produce Prescriptions. The Committee provides funding consistent with the budget request for VA facilities to implement local VA Produce Prescription programs and pursue strategies for patient education and outreach.

*Food Service Guidelines.*—The Committee supports the goal set forth in the National Strategy on Hunger, Nutrition, and Health to implement the Food Service Guidelines for Federal Facilities and encourages VA to use these guidelines across its food service facilities as soon as possible. The Department should provide, within 1 year of enactment of this act, a report to the Committees on Appropriations of both Houses of Congress that includes a baseline and final assessment of how VHA's food purchasing and food service practices align with the Food Service Guidelines, an evaluation of the implementation strategy including associated costs and customer feedback, and recommendations for changes to procurement or other laws that would facilitate ongoing implementation of the Food Service Guidelines.

*Whole Genome Sequencing Technologies.*—The Committee directs the Secretary to provide a report 1 year after the enactment of this act to the

Committees on Appropriations of both Houses of Con gress on the feasibility and advisability of utilizing whole genome sequencing technologies to conduct proactive sequencing for healthcare-associated infections to allow for faster interventions and better patient outcomes.

*Reducing Airborne Contaminants.*—The Committee recognizes that airborne bacteria and viruses are a source of risk to the health and well-being of veterans and staff at VA facilities. The Committee directs the Department to study the efficacy of UV technology platforms to address airborne contaminants in highly occupied and frequented spaces at VA facilities. The Committee directs VA to provide a report to the Committees on Appropriations of both Houses of Congress on the feasibility and advisability of utilizing such technologies no later than 1 year after enactment of this act.

*Wound Treatment.*—The Committee is encouraged by recent innovations in wound treatment technologies that could significantly reduce recovery times and lower treatment costs. The Committee encourages the Department to consider using these new technologies for wound treatment and directs the Department to provide a report to the Committees on Appropriations of both Houses of Congress, no later than 180 days after enactment of this act, on the feasibility and advisability of utilizing such technologies.

*Diabetes Center of Excellence.*—The Committee encourages VA to consider establishing a Diabetes Center of Excellence to focus on improving treatment and outcomes for Veterans with diabetes. The Center would conduct and support research, including health services research, and other activities to develop and disseminate best practices with respect to treatment (including modalities to induce remission or reversal) and prevention within VHA.

*Diabetic Foot Ulcers.*—The Committee commends VA's work to eliminate diabetes-related foot amputations and supports the Department's continued investment in programs used to treat veterans with diabetic foot ulcers. Further, the Committee encourages the Department to develop the Remote Temperature Monitoring Program, and to the extent practicable, expand its use of clinically validated remote monitoring technology. The Department is directed to provide to the Committees on Appropriations of both Houses of Congress a report identifying its plans to support expanding the Remote Temperature Monitoring Program, and any barriers to doing so, including staffing levels no later than 120 days after the enactment of this act.

*Service Dogs.*—The Department is directed to produce a report for the Committees on Appropriations of both Houses of Congress on the feasibility, advisability and cost of operating a program to provide grants to nonprofit organizations for the purpose of assisting qualified organizations in carrying out programs to provide service dogs to eligible veterans. The Department should provide the report to the Committees no later than 1 year after enactment of this act.

*Neurology Centers of Excellence.*—The Committee recognizes the increasing number of veterans affected by neurologic conditions, including but not limited to: epilepsy, headache, multiple sclerosis, and Parkinson's disease. The Neurology Centers of Excellence provide essential innovative clinical care, education, and research efforts focused on these conditions. The Committee encourages fur ther investment in, and collaboration between the Centers across disciplines and directs the Department to allocate $70,000,000 for the Neurology Centers of Excellence, including $25,000,000 for the Epilepsy Centers, $24,000,000 for the Headache Centers, $6,500,000 for the MS Centers, and $14,000,000 for the PADRECCs. The Committee also directs the Department of Veterans Affairs to report back the use and allocation of funds for each Center within 120 days of enactment.

*Dental Services.*—The Committee encourages the Department to offer veterans robust support to address their oral health needs using a whole-person approach, especially for veterans with additional medical needs including diabetes and heart disease. The Department is reminded of the report required in S. Rept. 118–43 requiring an assessment on the state of veterans' oral healthcare programs and these programs' ability to adequately serve the current and future oral health needs of veterans. The Committee looks forward to reviewing this report.

*Over-the-Counter [OTC] Hearing Aids.*—The Committee encourages the Department to improve access to OTC hearing aids for veterans receiving audiology treatment at VA facilities, as clinically appropriate.

*Medical Marijuana.*—The Committee recognizes that the Department of Justice's Drug Enforcement Agency has concurred with the Department of Health and Human Services' 2023 recommendation to reschedule cannabis in the Controlled Substances Act from its current placement in Schedule I to the less restrictive Schedule III. Should cannabis be rescheduled to a lower Schedule, VA should consider issuing guidance allowing VHA doctors and other personnel to discuss, recommend, and facilitate access to medical marijuana in States with state-legal medical marijuana programs to the extent allowable under Federal law.

*Dialysis Services.*—As the Department reaches the end of the current Nationwide Dialysis Services contract period, VA should ensure that care is not disrupted or diminished for the veterans currently receiving treatment for dialysis services. The Committee remains interested in value-based dialysis payment models and directs VA to submit the feasibility report, due December 2022, no later than 3 months after the date of the enactment of this act to the Committees on Appropriations of both Houses of Congress.

*Focused Ultrasound Treatment.*—The Committee remains interested in research in non-invasive neuromodulation and brain therapeutics technology and application development utilizing focused ultrasound. The Committee looks forward to receiving the report directed in S. Rept. 118–43 on the feasibility and advisability of utilizing non-invasive Focused Ultrasound Technologies.

*Peer Visitors.*—The Committee commends the Department's work with veterans living with limb loss, particularly its support of programs that engage amputees to become certified peer visitors for other veterans. The Committee encourages the Department to consider expanding its certified peer visitor program to meet the needs of all veterans.

*Limb Loss.*—The Department is directed to submit a report to the Committees on Appropriations of both Houses of Congress no later than 90 days after enactment of this act on the feasibility, ad visability, and cost of expanding amputee services and educational opportunities for individuals who have received osseointegration surgery and are using osseointegration prosthetic technology.

*Veteran Access.*—The Department is directed to provide to the Committees on Appropriations of both Houses of Congress monthly updates, by facility, on the average number of days it takes for veterans to receive appointments at VA and in the community, from file entry date to first scheduled and file entry date to appointment, beginning 30 days after enactment of this act.

*Hyperbaric Oxygen Therapy [HBOT].*—The Committee urges the Department to consider continuing to research the efficacy of HBOT as a treatment for post-traumatic stress disorder and traumatic brain injury, and should permit case-by-case referrals for veterans to receive HBOT in the community.

*MRI-Based Solutions.*—The Committee encourages VA to consider partnerships to utilize MRI based solutions to identify, stratify and manage the metabolic disease burden in the VA population.

*Essential Medical Devices.*—The Committee urges VA to purchase essential medical devices, like needles and syringes, from U.S. manufacturers or allied trading partners, to the extent possible. The Committee directs VA to submit a report to the Committees on Appropriations of both Houses of Congress no later than 90 days after enactment of this act on any utilization of Chinese made devices among the 96 medical countermeasures published by FDA in 2020. The report should also include specific recommendations on how VA can reduce its dependency on Chinese medical devices and increase the domestic supply chain, while safeguarding against shortages.

*Per- and Polyfluoroalkyl Substance Exposure.*—The Committee is encouraged by the Department's review of the National Academies of Sciences, Engineering and Medicine recommendations related to the utility of blood testing, and its collaboration with Federal and academic partners, to better understand the health implications of per- and polyfluoralkyl [PFAS] exposure related to military service. The Committee directs VA to provide a report within 180 days of enactment of this act to the Committees on Appropriations of both Houses of Congress on the outcomes of VA's Health Outcomes Military Exposures [HOME] intra-agency Technical Working Group [TWG] on improved clinical care decisions in terms of screening, care, and compensation policies for veterans impacted by PFAS exposure during their military service, including costs associated with the Secretary's decision. The report should include TWG's final course of action and feedback provided by key veteran stakeholders on VA's planned way forward.

*Securing Veterans Health Data.*—The Committee is concerned that adversaries can gain unfettered access to sensitive health data of veterans without their knowledge or consent. Therefore, the Committee directs the Secretary to provide a report, no later than 90 days after enactment of this act, to the Committees on Appropriations of both Houses of Congress regarding biotechnology equipment or services that the Department has purchased, or biotechnology agreements or contracts the Department has agreed to, over the last five fiscal years, and efforts made to safeguard health data as part of those contracts.

## Medical Community Care

| | |
|---|---|
| Appropriations, 2024 | $30,342,023,000 |
| Advance appropriations, 2025 | 20,328,000,000 |
| Budget estimate, advance appropriations, 2026 | 34,000,000,000 |
| Committee recommendation, advance appropriations, 202 | 34,000,000,000 |

## Program Description

The Medical Community Care account provides for medical services for eligible enrolled veterans and other beneficiaries that is purchased from and provided by non-Department of Veterans Affairs facilities and providers, including contract hospitals, State homes, and outpatient services.

## Committee Recommendation

In fiscal year 2024, the Committee provided an advance appropriation of $20,328,000,000 for fiscal year 2025 for the Medical Community Care account. In addition, VA estimates to use $9,770,646,000 from the Toxic Exposures Fund to support medical community care in fiscal year 2025, for a total of $30,098,646,000. The Committee recommendation also includes an advance appropriation of $34,000,000,000 for Medical Community Care for fiscal year 2026.

*Standard Episode of Care.*—Within 90 days of enactment, the Committee directs VA to perform an analysis of how the Department could better leverage its Third Party Administrators [TPA] to inform VA when an enrolled veteran has been stabilized and may require inpatient admission or a high-dollar procedure that could be accommodated via the direct care system. Further, the analysis should consider whether such follow on care could be separated from the emergency care Standard Episode of Care.

*Third Party Administrators [TPA] Accountability.*—The Department should ensure CCN providers are held to the same access standards as those required for the Department by also requiring CCN providers to report real-time appointment wait times as a condition of their participation in the CCN. The Department should establish care quality, training, and screening standards in line with those required of VA clinical staff that ensure veterans receive private care at facilities as good as or better than VA facilities. Additionally, the Department should require TPAs to submit metrics for appointment wait times per category of healthcare to the Secretary at an interval determined by the Secretary.

*Community Care Authorizations.*—The Committee is concerned about reports of veterans having authorizations for community care revoked. As such, the Department shall provide a report to the Committees on Appropriations of both Houses of Congress outlining in aggregate the number of instances where existing community care authorizations were rescinded and provide examples of the rationale by which such a decision would be made. The report shall be provided within 270 days of enactment of this act.

## Medical Support and Compliance

| | |
|---|---|
| Appropriations, 2024 | $10,750,000,000 |
| Advance appropriations, 2025 | 11,800,000,000 |
| Budget estimate, advance appropriations, 2026 | 12,700,000,000 |
| Committee recommendation, advance appropriations, 2026 | 12,700,000,000 |

## Program Description

The Medical Support and Compliance account provides funds for management, security, and administrative expenses within the VA healthcare system, in addition to providing costs associated with the operation of VA medical centers and clinics, VISN offices, and the VHA Central Office in Washington, DC. This appropriation also covers Chief of Staff and Facility Director operations, quality of care oversight, legal services, billing and coding activities, procurement, financial management, security, and human resource management.

## Committee Recommendation

In fiscal year 2024, the Committee provided an advance appropriation of $11,800,000,000 for fiscal year 2025 for the Medical Support and Compliance account.

The Committee recommendation includes an advance appropriation of $12,700,000,000 for Medical Support and Compliance for fiscal year 2026, consistent with the budget request.

*Drug Quality Scoring.*—The Committee encourages the Department to partner with the Uniformed Services University of the Health Sciences' Center for Health Services Research to evaluate the quality of drug products for essential medicines and develop a scoring tool for potential use in procurement.

**Medical Facilities**

| | |
|---|---|
| Appropriations, 2024 | $8,949,485,000 |
| Advance appropriations, 2025 | 9,400,000,000 |
| Budget estimate, 2025 | |
| Committee recommendation, 2025 | 3,000,000 |
| Budget estimate, advance appropriations, 2026 | 700,000,000 |
| Committee recommendation, advance appropriations, 2026, | 700,000,000 |

**Program Description**

The Medical Facilities account provides funds for the operation and maintenance of the VA healthcare system's vast capital infrastructure. This appropriation provides for costs associated with utilities, engineering, capital planning, leases, laundry, groundskeeping, housekeeping, facility repair, and property disposition and acquisition.

**Committee Recommendation**

In fiscal year 2024, the Committee provided an advance appropriation of $9,400,000,000 for fiscal year 2025 for the Medical Facilities account. The recommendation for fiscal year 2025 includes an additional $3,000,000 to support expansion of child care centers at VA medical facilities, which coupled with the advance appropriation provided for fiscal year 2025 provides the Department with total budget authority of $9,403,000,000.

The Committee recommendation also includes an advance appropriation of $9,700,000,000 for Medical Facilities for fiscal year 2026.

*Prohibition on Smoking.*—The Committee supports the Department's efforts to make VA facilities smoke-free and encourages full implementation and enforcement of VHA Directives 1085 and 1085.1, which prohibit the use of cigarettes, cigars, pipes, and any other combustion or heating of tobacco, and the use of any electronic nicotine delivery system, including electronic or e-cigarettes, vape pens, and e-cigars by any person on the premises of any VA facility.

*Water Management Policies.*—Given the concern related to levels of water use in hospitals and hospital laboratories, especially when dealing with water scarcity in States with constrained resources, VA is urged to consider water efficiency and zero water systems when feasible.

*Veteran Access to DoD Bases.*—The Committee notes that Congress greatly expanded base access to a range of veterans and their caregivers, including Medal of Honor and Purple Heart recipients. However, this expansion of base access is limited and not all veterans, survivors and caregivers may access DoD facilities. Within 180 days, the Committee directs the Department to provide estimates of the total number of veterans, survivors and caregivers and to report on the issues and challenges presented by expanding the Department's issuance of the Veteran Health Identification Card that the Department of Defense accepts for base access.

*VA Non-Recurring Maintenance.*—The Committee is concerned that repeated maintenance issues related to air conditioning outages and poor facility conditions can lead to interruption in veterans' care. The Department should make every effort to pursue timely, and permanent solutions to facility maintenance issues to avoid disruption to care. Further, the Committee directs VA to assess its facilities where there have been repeated instances of poor facility or maintenance conditions with relation to air conditioning or failures over the past 5 years and analyze the timelines in which such maintenance issues were fixed, and whether any additional issues arose following the maintenance fixes. Additionally, VA shall examine the data processes utilized to determine if the systems were adequately fixed and assess whether such processes could be improved. VA should focus this assessment on facilities that have had a high proportion of new veteran patients in recent years, to assess the impact of changing veteran demographics on VA infrastructure.

*Infrastructure Prioritization.*—The Committee is aware that several categories of healthcare are seeing high demand, and due to VA's limitation of physical space capacity to deliver these services, that care is being sent into the community. Within 270 days of enactment of this act, the Department is directed to report to the Committees on Appropriations of both Houses of Congress on the feasibility, advisability and costs of prioritizing construction projects to meet the need for high-demand, high-cost medical services where the Department lacks physical space capacity. The report should include how prioritizing these types of medical services into current processes can improve veterans access to VA direct care and should include projects broken down by project type, estimated cost, and whether the project is currently funded or unfunded.

## Medical and Prosthetic Research

| | |
|---|---|
| Appropriations, 2024 | $943,000,000 |
| Budget estimate, 2025 | 868,000,000 |
| Committee recommendation | 878,000,000 |

## Program Description

The Medical and Prosthetic Research account provides funds for medical, rehabilitative, and health services research. Medical research supports basic and clinical studies that advance knowledge leading to improvements in the prevention, diagnosis, and treatment of diseases and disabilities. Rehabilitation research focuses on rehabilitation engineering problems in the fields of prosthetics, orthotics, adaptive equipment for vehicles, sensory aids, and related areas. Health services research focuses on improving the effectiveness and economy of the delivery of health services.

## Committee Recommendation

The Committee recommends $878,000,000 for the Medical and Prosthetic Research account. This is $10,000,000 above the budget request. In addition, VA estimates to use $59,000,000 from the Toxic Exposures Fund to support medical research in fiscal year 2025, for a total of $937,000,000.

The Committee remains highly supportive of this program, and recognizes its importance both in improving healthcare services to veterans and recruiting and retaining high quality medical professionals in the Veterans Health Administration.

*Advanced Platform Technology [APT] Center.*—The Committee applauds VHA for progress being made to provide our Nation's veterans with new assistive and restorative technologies that address sensory, motor, or cognitive deficits, as well as limb loss. The APT Center has effectively utilized a partnership model to leverage local and national research expertise to drive progress in diverse areas including artificial lung development, wound healing, and neurally-connected sensory prosthesis. The bill provides continued support for these efforts.

*Optimizing Prosthetics for Veterans.*—The Committee encourages the Department to conduct and support research activities that test and develop novel methods, including artificial intelligence models, imaging, and motion capture technologies, for designing, fitting, and adjusting prosthetic technologies to improve prosthetics for veterans.

*Reducing Opioid Use Through Medical Marijuana.*—The Committee encourages VA to conduct a study on the relationship between treatment programs involving medical marijuana that are approved by States, the access of veterans to such programs, and a reduction in opioid use and abuse among veterans, as data is available to do so. Findings should be provided to the Committee on Appropriations of both Houses of Congress within 1 year of enactment of this act.

*Per- and Polyfluoroalkyl [PFAS] Exposure.*—The Committee continues to direct VA to collaborate with the Department of Defense PFAS Task Force, academia, and other health institutions and agencies to monitor research activities, results, and publications on the health effects from PFAS. The Secretary, in coordination with the Secretary of Defense, shall submit to Congress a report on this collaboration not later than 180 days following the enactment of this act.

*Women Veterans Health Research.*—The Committee encourages VA to advance and expand the inclusion of women veterans in clinical research conducted by the Department by utilizing the provisions outlined in Public Law 100–322 related to discovery and data collaboration between VA and its partnering entities.

*Aviator Cancers Examination Study.*—The Committee encourages the Secretary to coordinate with the National Academies of Sciences, Engineering, and Medicine to conduct a study on the prevalence and mortality of cancer among individuals who served as active-duty aircrew of fixed-wing aircraft in the U.S. Armed Forces. The study should leverage findings in existing literature, to include the Air Force and Department of Defense studies. The Secretary should report on any findings of this study within 3 months of its completion to the Committees on Appropriations of both Houses of Congress.

## Medical Care Cost Recovery Collections Medical Care Collection Fund

| | |
|---|---|
| Appropriations, 2024 | $3,991,000,000 |
| Budget estimate, 2025 | 4,632,000,000 |
| Committee recommendation | 4,632,000,000 |

**Medical Care Collection Fund–Revenues Applied**

| | |
|---|---|
| Appropriations, 2024 | ¥$3,991,000,000 |
| Budget estimate, 2025 | ¥4,632,000,000 |
| Committee recommendation | ¥4,632,000,000 |

**Program Description**

The Medical Care Collection Fund [MCCF] was established by the Balanced Budget Act of 1997 (Public Law 105–33). In fiscal year 2004, Public Law 108–199 allowed the Department of Veterans Affairs to deposit first-party and pharmacy co-payments; third-party insurance payments and enhanced-use collections; long-term care co-payments; Compensated Work Therapy Program collections; and Parking Program fees into the MCCF. The Secretary of Veterans Affairs has the authority to transfer funds from the MCCF to the Medical Services and Medical Community Care accounts.

**Committee Recommendation**

The Committee recommendation includes the authority to retain co-payments and third-party collections, estimated to total $4,632,000,000 in fiscal year 2025.

**National Cemetery Administration**

| | |
|---|---|
| Appropriations, 2024 | $480,000,000 |
| Budget estimate, 2025 | 495,000,000 |
| Committee recommendation | 495,000,000 |

**Administration Overview**

The National Cemetery Administration [NCA] was established in accordance with the National Cemeteries Act of 1973 (Public Law 93–43). It has a four-fold mission: to provide for the interment in any national cemetery of the remains of eligible deceased servicemembers and discharged veterans, together with their spouses and certain dependents, and permanently maintain their graves; to provide headstones for, and to mark graves of, eligible persons in national, State, and private cemeteries; to administer the grant program for aid to States in establishing, expanding, or improving State veterans cemeteries; and to administer the Presidential Memorial Certificate Program.

## Committee Recommendation

The Committee recommends $495,000,000 for the National Cemetery Administration. This is an increase of $15,000,000 above the fiscal year 2024 enacted level and equal to the budget request.

The Committee has included bill language to make available through September 30, 2026, up to 10 percent of the National Cemetery Administration appropriation.

*National Cemetery Administration Access.*—The Committee directs VA to work in collaboration with DoD to allow veterans and their families to gain access at cemeteries that are located on DoD installations. Access should be granted in a way that maintains security, while allowing appropriate access for grieving families, including those without military identification.

*Rural Initiative Cemetery Infrastructure.*—The Committee supports the National Cemetery Administration's efforts to construct additional infrastructure at its Rural Initiative national cemeteries. The Committee encourages the Department to work with State and local entities, including non-profit organizations, interested in constructing and donating VA-approved facilities to meet the location- specific needs of veterans, visitors, and volunteers. The Department is directed to report to the Committees on Appropriations of both Houses of Congress about the barriers that exist to implementing the ''Hershel 'Woody' Williams State Committal Shelter Pilot Program Act.''

## Departmental Administration

| | |
|---|---|
| Appropriations, 2024 | $10,677,361,000 |
| Budget estimate, 2025 | 10,796,133,000 |
| Committee recommendation | 10,860,146,000 |

## Administration Overview

Departmental Administration provides for the administration of veterans benefits through VBA, the executive direction of the Department, several top level supporting offices, and the Board of Veterans Appeals.

## Committee Recommendation

The Committee recommends $10,860,146,000 for Departmental Administration. The amount is composed of $457,000,000 for General Administration; $267,000,000 for the Board of Veterans Appeals; $6,276,693,000 for Information Technology Systems; $894,000,000 for the Veterans Electronic Health Record; $301,000,000 for the Office of Inspector General; $2,069,000,000 for Construction, Major Projects; $380,453,000 for Construction, Minor Projects; $150,000,000 for Grants for Construction of State Extended Care Facilities; and $65,000,000 for Grants for the Construction of State Veterans Cemeteries.

In addition, funds are available from the Recurring Expenses Transformational Fund [Transformational Fund]. Established in the Consolidated Appropriations Act of 2016 (Public Law 114–113), the Transformational Fund consists of unobligated balances of expired discretionary appropriations, following the fifth fiscal year after the last fiscal year funds were originally available. These funds shall be available until expended for facilities infrastructure improvements, including nonrecurring maintenance, at existing VA hospitals and clinics, and for information technology systems improvements and sustainment. For fiscal year 2025, VA proposed to allocate $307,000,000 of the anticipated Transformational Fund balance for minor construction projects.

## General Administration (Including Transfer of Funds)

| | |
|---|---|
| Appropriations, 2024 | $475,000,000 |
| Budget estimate, 2025 | 457,000,000 |
| Committee recommendation | 457,000,000 |

## Program Description

The General Administration account provides funding for the Office of the Secretary, six assistant secretaries, and three independent staff offices.

## Committee Recommendation

The Committee recommends $457,000,000 for General Administration, equal to the budget request. The Committee has included bill language to make available through September 30, 2026, up to 10 percent of the General Administration appropriation.

The Committee provides funding for General Administration in the amounts specified below:

[In thousands of dollars]

| Office 1 | Fiscal year 2025 budget request | Committee recommendation |
|---|---|---|
| Office of the Secretary .......................................................... | 18,159 | 18,159 |
| Office of General Counsel ..................................................... | 142,310 | 142,310 |
| Office of Management ........................................................... | 81,232 | 81,232 |
| Office of Human Resources & Administration/Operations, Security & Preparedness ......................................... | 121,727 | 121,727 |
| Office of Enterprise Integration ........................................... | 37,818 | 37,818 |
| Office of Public & Intergovernmental Affairs ........................ | 16,298 | 16,298 |
| Office of Congressional & Legislative Affairs ...................... | 9,895 | 9,895 |
| Office of Accountability & Whistleblower Protection .......... | 29,561 | 29,561 |
| Total ....................................................................................... | 457,000 | 457,000 |

[1]The Office of Acquisition, Logistics & Construction and the Veterans Experience Office are funded solely with reimbursable authority.

The Secretary may alter these allocations if the Committees have been notified and written approval is provided.

*Unobligated Balances of Expired Discretionary Funds.*—The Committee directs the Secretary to submit to the Committees on Appropriations of both Houses of Congress quarterly reports detailing all unobligated balances of expired discretionary funds by fiscal year.

*Access to High-Quality Affordable Childcare for Federal Employees.*—The Committee applauds the Department for increasing the childcare subsidy for employees through the VA Childcare Subsidy Program [CCSP], but is concerned that the subsidy the Department is providing employees may not have kept pace with the rising costs of childcare. Within 180 days of enactment, the Committee directs the Department to provide a report to the Committees on Appropriations of both Houses of Congress on the number of employees using the subsidy in each tier of the CCSP, the percentage of childcare costs the CCSP covers for those employees that opt to use the program, a breakdown of the percentage of the funds spent on CCSP for center-based care, home care, and or before/after care respectively, how many eligible employees forgo the subsidy and how the maximum monthly subsidy compares to the average cost of childcare in high cost and rural markets.

*Advertising Contracts.*—The Committee understands that, as the largest advertiser in the United States, the Federal Government should work to ensure fair access to its advertising contracts for small disadvantaged businesses and businesses owned by minorities and women. The Secretary is directed to provide a report to the Committees on Appropriations of both Houses of Congress within 180 days of enactment of this act with the following information: expenditures for fiscal year 2024 and expected expenditures for fiscal year 2025 for all contracts for advertising services; and contracts for the advertising services of socially and economically disadvantaged small businesses concerns (as defined in section 8(a)(4) of the Small Business Act (15 U.S.C. 637 (a)(4))); and women-and minority-owned businesses.

*Consolidated Mail Operations.*—The Secretary may be able to improve services to veterans and their families, increase efficiency, and reduce operational costs by investing in modern information technology systems and centralizing the Department's mail management operations, with the exception of mailing prescription drugs, to allow VA to manage its correspondence on an enterprise level. The Secretary should consider centralizing mail operations to operate on a shared enterprise service basis so as to drive production efficiency and mail volumes capable of taking advantage of automation discounts offered by USPS. The Secretary is directed toprovide a report to the Committees on Appropriations of both Houses of Congress, no later than 180 days after enactment of this act, detailing the potential savings, benefits, and improvements in service to veterans of such an effort.

*VA Fellowship Program.*—The Department is directed to provide a report to the Committees on Appropriations of both Houses of Congress on the advisability, feasibility, and cost of establishing a permanent, structured, and centralized VA Fellowship Program that includes the selection and training of promising candidates, to include thorough grounding in all of the Department's major functions; addresses candidate expenses for moving to the Washington, DC area and back to their home station or a follow-on assignment; and integrates the fellowship with other professional development programs for emerging leaders, as appropriate.

*Voting-Related Activities.*—The Committee notes that the Department has not yet provided the briefing required under this heading in the explanatory statement accompanying Public Law 118–42. As such, the Committee directs the Secretary to provide to the Committees on Appropriations of both Houses of Congress a briefing, within 30 days after enactment of this act, regarding any strategic plans developed by the Department since January 20, 2021 outlining ways for the Department to promote voter registration and voter participation.

## Board of Veterans Appeals

| | |
|---|---|
| Appropriations, 2024 | $287,000,000 |
| Budget estimate, 2025 | 267,000,000 |
| Committee recommendation | 267,000,000 |

## Program Description

As set forth in section 7101(a) of title 38 United States Code, the Board of Veterans Appeals [BVA] is responsible for making final decisions on claims for veterans benefits presented to the Board for appellate review. The vast majority of the Board's workload derives from benefit claims initiated by the Veterans Benefits Administration's Regional Offices. The appellate process has multiple steps, most of which occur at the local Regional Office level. If a veteran is not satisfied with the Regional Office determination, he or she may appeal to the Board for a final agency decision. The Board adjudicates appeals covering all areas of veterans benefits, including service connection, increased disability ratings, total disability ratings, pensions, insurance benefits, educational benefits, home loan guaranties, vocational rehabilitation, waivers of indebtedness, fee basis medical care, and dependency and indemnity compensation.

## Committee Recommendation

The Committee recommends $267,000,000 for the Board of Veterans Appeals, equal to the budget request. The Committee has included bill language to make available through September 30, 2026, up to 10 percent of BVA's appropriation. In addition, VA estimates to use $19,000,000 from the Toxic Exposures Fund to support BVA activities related to toxic exposures in fiscal year 2025.

*Appeals Backlog.*—The Committee is concerned with the large backlog of cases at the Board of Veterans Appeals. The Departmentis directed to provide a report to the Committees on Appropriations of both Houses of Congress no later than 90 days after enactment of this act about its plans to eliminate the backlog.

*Staffing.*—The Committee supports the Board of Veterans Appeals' goal to process 118,000 appeals in fiscal year 2025. To accomplish this goal, the Committee encourages the Board to use all resources available to hire additional staff and judges to keep pace with the expected increased workload of appeals from toxic-exposed veterans.

## Information Technology Systems

| | |
|---|---|
| Appropriations, 2024 | $6,401,000,000 |
| Budget estimate, 2025 | 6,231,680,270 |
| Committee recommendation | 6,276,693,000 |

## Program Description

The Information Technology [IT] Systems appropriation, along with reimbursements, funds the costs of all IT staff salaries and expenses, the operations and maintenance of all existing information technology systems, and the development of new projects and programs designed to improve the delivery of service to veterans. This appropriation also funds the costs associated with the Office of Information and Technology which oversees the functions highlighted above.

## Committee Recommendation

The Committee recommends $6,276,693,000 for the Information Technology Systems account. This amount is $45,013,000 above the budget request. The Committee recommendation includes $1,695,247,816 for staff salaries and expenses, $4,575,983,911 for operation and maintenance of existing programs, including $118,900,000 for activations, and $5,461,273 for program development. In addition, VA estimates to use $1,364,000,000 from the Toxic Exposures Fund to support the information technology systems activities related to toxic exposures in fiscal year 2025.

The Committee has appropriated the Information Technology Systems account as three subaccounts. This funding structure enhances the Committee's ability to ensure funds are executed in a manner consistent with the Department's budget submission. The Committee has provided sufficient flexibility within the sub-accounts by way of authorized carryover amounts and reprogramming authority to give the Office of Information Technology as much flexibility as possible to accomplish its mission and goals, while ensuring proper accountability and oversight.

*Artificial Intelligence [AI] Initiative.*—The Department is encouraged to utilize AI to improve the Veteran experience in ways that align with the Executive Order 14110, Safe, Secure, and Trustworthy Development and Use of Artificial Intelligence. The Committee further believes the Department would benefit from the establishment of an AI Governing Council. Additionally, the Department should consider leveraging commercially-available technology to improve efficiency, hiring, and onboarding processes. The Department is directed to provide a report to the Committees on Appropriations of both Houses of Congress no later than 180 days after enactment of this act, on the areas in which the Department identifies a positive use case for AI in ways that support the VA workforce. The report should include ways to reduce the administrative burden of clinicians and administrative staff, and to improve interfaces which reduce the difficulty of scheduling appointments, receiving reimbursement of covered veteran expenses, the billing of services, and other efforts that serve veterans.

*Accessibility.*—The Committee believes the Department would benefit from the establishment of an Advisory Committee on Equal Access to ensure the Department's compliance with disability laws. The Committee also recognizes a continued need for the Department to improve the accessibility of its technology for people with disabilities. The General Services Administration's [GSA] fiscal year 2023 Government-wide Section 508

Assessment, and a recent Office of Inspector General Report, both highlight the need for the Department to improve accessibility. GSA's assessment of the Department did not include data on the accessibility of VA's internal intranet, public electronic documents, or videos. The Committee encourages the Department to provide GSA with testing outcomes for all of its information and communication technology for the next annual GSA assessment of Section 508.

## Veterans Electronic Health Record

| | |
|---|---|
| Appropriations, 2024 | $1,334,142,000 |
| Budget estimate, 2025 | 894,000,000 |
| Committee recommendation | 894,000,000 |

### Program Description

The Veterans Electronic Health Record [EHR] account funds all activities related to the acquisition, implementation, preparation, development, interface, management, rollout, and maintenance of a new EHR. The EHR solution and implementation will include program management; an enterprise-wide EHR system; change management; training; testing; deployment services; sustainment; and other solutions encompassing the entire range of EHR requirements.

### Committee Recommendation

The Committee recommends $894,000,000 for the Veterans EHR, equal to the budget request. The Committee remains supportive of replacing VA's current EHR with the same system being acquired by DoD, and supports the Department's current effort to ''reset'' the initiative and hold future rollouts until performance in the sites currently using the new EHR has improved, but urges the Department to restart deployments as soon as is safe.

Given the schedules and costs associated with any acquisition of this size and magnitude, it is important for the Department to provide regular implementation updates. Therefore, the bill continues the strict quarterly reporting requirements of obligations, expenditures, and any upcoming deployment schedule by facility. The Department is directed to continue quarterly briefings to review timelines, performance milestones, costs, implementation, and change management progress. In addition, the Committee includes a statutory proviso that makes 25 percent of funding contingent upon the Secretary (1) providing notification of each instance when the contractor supporting acquisition has failed to meet its obligations for system availability; (2) a plan to standardize the system for deployment across the enterprise; and (3) providing periodic updates, beginning 30 days after enactment of this act, on the progress of upgrades to the pharmacy and billing applications.

The Committee continues to direct GAO to perform quarterly performance reviews of the VA EHR deployment to keep the Committees on Appropriations and Veteran's Affairs of both Houses of Congress apprised of VA's progress.

*Medical Image Exchange*.—The Committee understands that there are VA facilities that have deployed a modern software platform that supports bi-directional exchange of medical images, diagnostic reports, and other clinical data between VA facilities, community providers, and Federal partners. VA is encouraged to expand this software's connections to Indian and Tribal health clinics and work with DoD to expedite connections with military treatment facilities. Further, VA should consider making this software platform available in the new Electronic Health Record system, the VA Enterprise Cloud and the anticipated national Enterprise Imaging initiative to build on the care quality, efficiency and productivity gains.

## Office of Inspector General

| | |
|---|---|
| Appropriations, 2024 | $296,000,000 |
| Budget estimate, 2025 | 296,000,000 |
| Committee recommendation | 301,000,000 |

## Program Description

The Office of Inspector General [OIG] was established by the Inspector General Act of 1978 and is responsible for the audit, investigation, and inspection of all Department of Veterans Affairs programs and operations.

## Committee Recommendation

The Committee recommends $301,000,000 for OIG, $5,000,000 above the budget request. The Committee has included bill language to make available through September 30, 2026, up to 10 percent of the OIG appropriation. Given the importance of the OIG work related to implementation of PACT Act (Public Law 117–168), the Committee includes $5,000,000 in funding for this work.

OIG continues to sustain an extremely high performance level, and the Committee continues to support OIG's essential oversight of VA's programs and operations, and maintains a provision to support timely access to any records, documents, or other materials available to the Department. Similarly, the Committee strongly encourages OIG to undertake and complete investigations in a timely manner, including sharing information with the Department, the Department of Justice, or other entities as appropriate.

## Construction, Major Projects

| | |
|---|---|
| Appropriations, 2024 | $961,219,000 |
| Budget estimate, 2025 | 2,069,000,000 |
| Committee recommendation | 2,069,000,000 |

## Program Description

The Construction, Major Projects account provides funds for constructing, altering, extending, and improving any of the facilities (including parking projects) under the jurisdiction or for the use of VA, including planning, architectural and engineering services, needs assessment, and site acquisition where the estimated cost of a project is more than the amount set forth in 38 U.S.C. 8104(a)(3)(A). Proceeds realized from enhanced use lease activities may be deposited into the Construction, Major Projects and Construction, Minor Projects accounts.

## Committee Recommendation

The Committee recommends an appropriation of $2,069,000,000 for the construction of major projects. This is equal to the budget request.

The following table reflects the President's budget request for major construction projects and the corresponding Committee recommendations.

Major Construction Projects

[In thousands of dollars]

| Location and description | Fiscal year 2025 budget request | Committee recommendation |
|---|---|---|
| Veterans Health Administration [VHA]: | | |
| West Los Angeles, CA: Build New Critical Care Center, Central Utility Plant, Demolition and Renovations to Building 500 ...................................... | 1,344,810 | 1,344,810 |
| Dallas, TX: Clinical Expansion for Mental Health, Expansion of Parking Facilities | | |
| and Land Acquisition ................................................ | 13,300 | 13,300 |
| Advance Planning and Design Fund—Various Stations ............................................. | 278,409 | 278,409 |
| Asbestos—Various Stations ...................................... | 1,000 | 1,000 |
| Construction & Facilities Management Staff—Various Stations ................................ | 213,000 | 213,000 |
| Hazardous Waste—Various Stations ........................ | 1,000 | 1,000 |
| Claims Analysis—Various Stations ........................... | 500 | 500 |
| Non-Departmental Federal Entity Project Management Support—Various Stations .... | 134,481 | 134,481 |
| Total, VHA ............................................................... | 1,986,500 | 1,986,500 |
| National Cemetery Administration [NCA]: | | |
| Denver, CO: Fort Logan National Cemetery—Gravesite Expansion ............................. | 45,000 | 45,000 |
| Advance Planning and Design Fund—Various Stations ............................................. | 30,000 | 30,000 |
| Total, NCA ............................................................... | 75,000 | 75,000 |
| General Administration/Staff Offices | | |
| Department Advance Planning and Design Fund for Major Construction ................... | 7,500 | 7,500 |
| Total Construction, Major Projects ........................... | 2,069,000 | 2,069,000 |

*Vulnerable Infrastructure.*—The Committee continues to encourage the Department to evaluate its infrastructure and construction needs taking into account the stress that recent historic weather trends have placed on aging or inadequate buildings and infrastructure. The Committee looks forward to receiving a report as requested in S. Rept. 118–43 on extreme weather preparation.

## Construction, Minor Projects

| | |
|---|---|
| Appropriations, 2024 | $692,000,000 |
| Budget estimate, 2025 | 380,453,000 |
| Committee recommendation | 380,453,000 |

## Program Description

The Construction, Minor Projects account provides for constructing, altering, extending, and improving any of the facilities (including parking) under the jurisdiction or for the use of VA, including planning, assessment of needs, architectural and engineering services, and site acquisition, where the estimated cost of a project is equal to or less than $30,000,000. The Veterans Millennium Health Care and Benefits Act (Public Law 106–117) gave VA authority to make capital contributions from minor construction in enhanced-use leases. Proceeds realized from enhanced-use lease activities may be deposited into the Construction, Major Projects and Construction, Minor Projects accounts.

## Committee Recommendation

The Committee recommends $380,453,000 for minor construction. This is equal to the budget request.

The recommendation includes $64,453,000 for the Veterans Health Administration, $174,075,000 for the National Cemetery Administration, $46,080,000 for the Veterans Benefits Administration, and $95,845,000 for staff offices and the Office of Information and Technology. In addition, the Committee supports the Department's allocation of $307,000,000 from the Transformational Fund to support the construction of minor projects bringing the Department's total resources for Minor Construction to $687,453,000. The Department is directed to provide an expenditure plan to the Committees on

Appropriations of both Houses of Congress no later than 30 days after enactment of this act for the amount appropriated for minor construction.

*Permanent Residential Rehabilitation Treatment Unit.*—The Committee recognizes the importance of long-term residential rehabilitation treatment unit beds for veterans suffering from substance use disorders, particularly in rural areas. The Department is encouraged to include such projects in future budgets.

## Grants for Construction of State Extended Care Facilities

| | |
|---|---|
| Appropriations, 2024 | $171,000,000 |
| Budget estimate, 2025 | 141,000,000 |
| Committee recommendation | 150,000,000 |

## Program Description

This account is used to provide grants to assist States in acquiring or constructing State home facilities for furnishing domiciliary or nursing home care to veterans, and to expand, remodel, or alter existing buildings for furnishing domiciliary, nursing home, or hospital care to veterans in State homes. The grant may not exceed 65 percent of the total cost of the project. Public Law 102–585 granted permanent authority for this program, and Public Law 106–117 provided greater specificity in directing VA to prescribe regulations for the number of beds for which grant assistance may be furnished. This program has been a successful partnership between States and VA in meeting the long-term care needs of elderly veterans for decades.

## Committee Recommendation

The Committee recommends $150,000,000 for Grants for Con struction of State Extended Care Facilities. This is $9,000,000 above the budget request.

*Grant Eligibility.*—The Secretary is urged to consider additional factors when reviewing and prioritizing grant applications, including whether a State has only one State Extended Care facility, the geographic diversity among grant recipients, whether a facility has an industry-standard Facility Condition

Index maximum rating of poor, and whether veterans would have access to other providers who can appropriately meet their needs.

**Grants for Construction of Veterans Cemeteries**

| | |
|---|---|
| Appropriations, 2024 | $60,000,000 |
| Budget estimate, 2025 | 60,000,000 |
| Committee recommendation | 65,000,000 |

**Program Description**

Public Law 105–368 amended title 38 U.S.C. 2408 and established authority to provide aid to States for establishment, expansion, and improvement of State veterans cemeteries, which are operated and permanently maintained by the States. This statutory change increased the maximum Federal share from 50 percent to 100 percent in order to fund construction costs and initial equipment expenses when the cemetery is established. States remain responsible for providing the land and for paying all costs related to operation and maintenance of the cemeteries, including the costs for subsequent equipment purchases.

**Committee Recommendation**

The Committee recommends $65,000,000 for Grants for Construction of State Veterans Cemeteries. This is $5,000,000 above the budget request.

*Veterans Cemetery Grants Program [VCGP].*—The Committee supports NCA's efforts to support States as they expand burial opportunities, especially for lower density veteran populations. In order to ensure that these infrastructure projects are meeting the needs of veterans, volunteers, and visitors, the Committee encourages VA to give appropriate consideration to States without a VCGP grant before expanding existing cemeteries.

## Cost of War Toxic Exposures Fund

| | |
|---|---|
| Appropriations, 2024[1] | $20,268,000,000 |
| Appropriations, 2025[1] | 24,455,000,000 |

[1]Funds for fiscal year 2024 and 2025 were appropriated in the Fiscal Responsibility Act of 2023 (Public Law 118–5).

## Program Overview

The Cost of War Toxic Exposures Fund [Toxic Exposures Fund] was established in order to fully fund the new costs related to providing veterans and their families the benefits and care associated with eligibility expansion included in the Honoring Our PACT Act of 2022 (Public Law 117–168) [PACT Act].

The Department may allocate funds from the Cost of War Toxic Exposures Fund for investment in the delivery of veterans healthcare associated with exposure to environmental hazards in service, expenses incident to the delivery of veterans healthcare and benefits associated with exposure to environmental hazards in service, and medical and other research related to exposures to environmental hazards in service.

The Committee supports the appropriations provided in the Fiscal Responsibility Act of 2023 (Public Law 118–5), and includes a provision requiring quarterly reporting on the status of the Toxic Exposures Fund. Further, the Committee continues to expect a consolidated section in budget requests outlining the proposed spending from the Toxic Exposures Fund, consistent with requirements in Public Law 117–168.

*Kosovo Veterans Toxic Exposures.*—As the Department undergoes rulemaking for Public Law 117–168, the Department should consider adding Kosovo to the list of covered countries for the purposes of determining presumptive eligibility, specifically service in the NATO Kosovo Force [KFOR] since June 1999, as ‘‘covered service.’’

**Administrative Provisions**

(INCLUDING TRANSFERS AND RESCISSIONS OF FUNDS)

Sec. 201. The Committee includes a provision which outlines transfer authority and responsibilities for the Veterans Benefits Administration.

Sec. 202. The Committee includes a provision which outlines transfer authority and responsibilities for the Veterans Health Administration.

Sec. 203. The Committee includes a provision which outlines the use of funds appropriated for salaries and expenses.

Sec. 204. The Committee includes a provision mandating that only construction funds may be used for land procurement or the construction of any new hospital or home.

Sec. 205. The Committee includes a provision allowing for reimbursements to the Medical Services account.

Sec. 206. The Committee includes a provision allowing for payments of prior year obligations.

Sec. 207. The Committee includes a provision which allows for the use of funds for prior year obligations.

Sec. 208. The Committee includes a provision which allows for payments from the National Service Life Insurance Fund.

Sec. 209. The Committee includes a provision which outlines the use of funds from enhanced-use lease proceeds.

Sec. 210. The Committee includes a provision which provides funds for the Office of Resolution Management, Diversity and Inclusion, the Office of Employment Discrimination Complaint Adju dication, and the Alternate Dispute Resolution function within the Office of Human Resources and Administration.

Sec. 211. The Committee includes a provision which requires disclosure of third-party reimbursement information.

Sec. 212. The Committee includes a provision which allows for the transfer of revenue derived from enhanced-use leases into the construction accounts.

Sec. 213. The Committee includes a provision which outlines authorized uses for Medical Services account funds.

Sec. 214. The Committee includes a provision which allows funds in the Medical Care Collection Fund to be transferred into the Medical Services and Medical Community Care accounts.

Sec. 215. The Committee includes a provision which allows eligible veterans in the State of Alaska to obtain medical care services.

Sec. 216. The Committee includes a provision which allows for the transfer of funds into the construction accounts.

Sec. 217. The Committee includes a provision requiring the Secretary of Veterans Affairs to submit quarterly financial reports.

Sec. 218. The Committee includes a provision outlining transfer authority for the Information Technology Systems account.

Sec. 219. The Committee includes a provision allowing for the transfer of funds from certain accounts to the Joint Department of Defense/Department of Veterans Affairs Medical Facility Demonstration Fund, as authorized by Public Law 111–84.

Sec. 220. The Committee includes a provision allowing for the transfer of funds from certain advance appropriation accounts to the Joint Department of Defense/Department of Veterans Affairs Medical Facility Demonstration Fund, as authorized by Public Law 111–84.

Sec. 221. The Committee includes a provision allowing for the transfer of certain funds deposited in the Medical Care Collections Fund to the Joint Department of Defense/Department of Veterans Affairs Medical Facility Demonstration Fund, as authorized by Public Law 111–84.

Sec. 222. The Committee includes a provision directing a minimum of $15,000,000 be transferred from Medical Services, Medical Support and Compliance, and Medical Facilities to the Department of Defense/Department of Veterans Affairs Health Care Sharing Incentive Fund, as authorized by section 8111 of title 38, United States Code.

Sec. 223. The Committee includes a provision prohibiting funds available to the Department in this or any other act from being used to replace the current system by which VISNs select and contract for diabetes monitoring supplies and equipment.

Sec. 224. The Committee includes a provision requiring notification of all bid savings for major construction projects.

Sec. 225. The Committee includes a provision restricting scope increases for major construction projects above that specified in the original project justification.

Sec. 226. The Committee includes a provision requiring the Department to submit reports relating to the Veterans Benefits Administration on claims processing at Regional Offices.

Sec. 227. The Committee includes a provision requiring VA to notify the Committee 15 days prior to any organizational changes within VA of 25 or more FTE.

Sec. 228. The Committee includes a provision requiring the Secretary to report to the Committees each quarter about any single national outreach and awareness marketing campaign exceeding $1,000,000.

Sec. 229. The Committee includes a provision permitting the transfer to the Medical Services account of fiscal year discretionary 2025 appropriated funds.

Sec. 230. The Committee includes a provision permitting the transfer of funds between GOE, VBA and BVA.

Sec. 231. The Committee includes a provision prohibiting the reprogramming of funds in excess of $7,000,000 among major construction projects or programs.

Sec. 232. The Committee includes a provision mandating certain professional standards for the veterans crisis hotline.

Sec. 233. The Committee includes a provision requiring VA to use the mammography screening guidelines announced by the Secretary on May 10, 2017.

Sec. 234. The Committee includes a provision allowing the use of Medical Services funding for assisted reproductive technology treatment and adoption reimbursement for veterans and their spouses if the veteran has a service-connected disability that results in being unable to procreate without such fertility treatment.

Sec. 235. The Committee includes a provision prohibiting any funds to be used to contract out any functions performed by more than 10 employees without a fair competition process.

Sec. 236. The Committee includes a provision pertaining to Native Hawaiian small businesses.

Sec. 237. The Committee includes a provision directing the discontinuation of the usage of Social Security numbers within VA.

Sec. 238. The Committee includes a provision pertaining to the certification of marriage and family therapists.

Sec. 239. The Committee includes a provision prohibiting the transfer of funds from the Filipino Veterans Equity Compensation Fund to any other VA account.

Sec. 240. The Committee includes a provision regarding a child care program.

Sec. 241. The Committee includes a provision prohibiting funds to be used to restrict an individual's ability to speak with a Member of Congress or his or her staff.

Sec. 242. The Committee includes a provision requiring certain data to be included in the budget justifications for the Construction, Major account.

Sec. 243. The Committee includes a provision prohibiting the use of funds to deny the Inspector General timely access to information, unless a provision of law expressly refers to the Inspector General and expressly limits such access.

Sec. 244. The Committee includes a provision prohibiting funding from being used in a manner that would increase wait times for veterans at medical facilities.

Sec. 245. The Committee includes a provision prohibiting the use of funds in fiscal year 2025 to convert any program which received specific purpose funds in fiscal year 2024 to a general purpose-funded program without the approval of the Committees on Appropriations of both Houses of Congress at least 30 days prior to any such action.

Sec. 246. The Committee includes a reference to a provision in the 2017 appropriations act identifying information which may be used to verify the status of coastwise merchant seamen who served during World War II for the purposes of eligibility for medals, ribbons, or other military decorations.

Sec. 247. The Committee includes a provision addressing animal research at the Department of Veterans Affairs.

Sec. 248. The Committee includes a provision ensuring particular ratios of veterans to FTE position within any VA program of rehabilitation.

Sec. 249. The Committee includes a provision to allow fiscal year 2025 and 2026 ''Medical Community Care'' funds to be used to cover obligations that otherwise would be paid by the Veterans Choice Fund, if necessary.

Sec. 250. The Committee includes a provision to allow obligations and expenditures applicable to the ''Medical Services'' account in fiscal years 2017 through 2019 for aid to State homes to remain in the ''Medical Community Care'' account.

Sec. 251. The Committee includes a provision specifying an amount from the four medical care accounts for gender-specific care and programmatic efforts to deliver care for women veterans.

Sec. 252. The Committee includes a provision allocating funds from the ''Recurring Expenses Transformational Fund'' for specific purposes.

Sec. 253. The Committee includes a provision requiring quarterly reports on obligations from the ''Cost of War Toxic Exposures Fund.''

Sec. 254. The Committee includes a provision to address contributions from other Federal agencies to VA Non-Profit Corporations for research.

Sec. 255. The Committee includes a provision to restrict funds from being used to close medical facilities.

Sec. 256. The Committee includes a provision to allow use of unobligated balances to support construction projects in the CHIP–IN program.

Sec. 257. The Committee includes a provision to rescind previously appropriated funding.

Sec. 258. The Committee includes a provision prohibiting the use of funds to administer, implement, or enforce the final rule relating to ''Change in Rates VA Pays for Special Modes of Transportation''.

Sec. 259. The Committee includes a provision prohibiting the use of funds for certain information technology equipment and software.

Sec. 260. The Committee includes a provision related to payment of awards and incentive fees to contractors with unsatisfactory performance.

## Title III: Related Agencies

### American Battle Monuments Commission

#### *Overview*

The American Battle Monuments Commission [ABMC] was established by Congress in 1923 and is responsible for the following: designing, constructing, operating, and maintaining permanent American cemeteries in foreign countries; establishing and maintaining U.S. military memorials, monuments, and markers where American Armed Forces have served overseas since April 6, 1917, the date of the United States entry into World War I, and within the United States when directed by public law; and controlling the design and construction of permanent U.S. military monuments and markers by other U.S. citizens and organizations, both public and private, and encouraging their maintenance. ABMC administers, operates, and maintains 26 permanent American military cemeteries and 32 Federal memorials, monuments, and markers, located in 17 foreign countries, the U.S. Commonwealth of the Northern Mariana Islands, the Midway Atoll, the British Dependency of Gibraltar, and the United States of America.

### Salaries and Expenses

| | |
|---|---|
| Appropriations, 2024 | $158,630,000 |
| Budget estimate, 2025 | 89,520,000 |
| Committee recommendation | 94,520,000 |

### Committee Recommendation

The Committee recommends $94,520,000 for the Salaries and Expenses account. This amount is $5,000,000 above the budget request.

### Foreign Currency Fluctuations

The Committee includes in the accompanying act, as proposed by the administration, such sums as necessary for the Foreign Currency Fluctuations account. Funding the account in this manner allows the Commission to maintain cemeteries regardless of the volatility of foreign currency fluctuations.

## United States Court of Appeals for Veterans Claims

***Overview***

The United States Court of Appeals for Veterans Claims was established by the Veterans' Judicial Review Act of 1988 (Public Law 100–687). The Court is an independent judicial tribunal with exclusive jurisdiction to review decisions of the Board of Veterans Appeals. It has the authority to decide all relevant questions of law; interpret constitutional, statutory, and regulatory provisions; and determine the meaning or applicability of the terms of an action by the Secretary of Veterans Affairs. It is authorized to compel action by the Secretary. It is authorized to hold unconstitutional or otherwise unlawful and set aside decisions, findings, conclusions, rules, and regulations issued or adopted by the Secretary of Veterans Affairs, the Board of Veterans Appeals, or the Chairman of the Board that are found to be arbitrary or capricious. The Court's principal office location is Washington, DC; however, it is a national court, empowered to sit anywhere in the United States.

## Salaries and Expenses

| | |
|---|---|
| Appropriations, 2024 | $47,200,000 |
| Budget estimate, 2025 | 47,300,000 |
| Committee recommendation | 47,300,000 |

## Committee Recommendation

The Committee recommends $47,300,000 for the U.S. Court of Appeals for Veterans Claims. This amount is $100,000 above the fiscal year 2024 enacted level and equal to the budget request.

*Judicial Expansion.*—The Committee supports efforts by the Court to expand its judicial bench in order to reduce adjudication times. To allow necessary flexibility for the Court to adapt to uncertain timelines and outcomes associated with efforts to increase authorization of judges, and to ensure efficient allocation of resources, the Committee provides 2-year funding limited to the amounts necessary to fund two additional judges and their support staff and services.

## Department of Defense—Civil Cemeterial Expenses, Army

### *Overview*

The Secretary of the Army is responsible for the administration, operation, and maintenance of Arlington National Cemetery [ANC] and the Soldiers' and Airmen's Home National Cemetery. In addition to its principal function as a national cemetery, Arlington hosts more than 3,000 public wreath laying ceremonies, and approximately 3 million visitors annually.

## Salaries and Expenses

| | |
|---|---|
| Appropriations, 2024 | $99,880,000 |
| Budget estimate, 2025 | 105,514,000 |
| Committee recommendation | 105,514,000 |

## Committee Recommendation

The Committee recommends $105,514,000 for the Salaries and Expenses account. This amount is $5,634,000 above the fiscal year 2024 enacted level and equal to the budget request.

ANC historically performs over 7,000 burial services each year for veterans and family members and an average of 27–30 each weekday. The Committee remains unwavering in its support for the Cemetery and the successful completion of the Cemetery's truly unique and honored mission.

*Virtual Tours.*—The Committee recognizes that technology, such as interactive tools and virtual tours, provides greater public exposure to the Cemetery and its mission, and encourages ANC to explore opportunities to leverage such options, to include reviewing applicability, cost, and potential limitations. The Committee directs ANC to provide a briefing on its findings no later than 90 days after enactment of this act.

## Construction

| | | |
|---|---|---|
| Appropriations, 2024 | | $88,600,000 |
| Budget estimate, 2025 | | 42,000,000 |
| Committee recommendation | 4 | 2,000,000 |

## Committee Recommendation

The Committee recommends $42,000,000 for the Construction account. This amount is equal to the budget request and includes funding for design of Memorial Avenue improvements and additional internment and columbarium expansions. Along with the Southern Expansion project, these efforts will extend the service life of the cemetery into the 2060s.

## Armed Forces Retirement Home Trust Fund

### *Overview*

The Armed Forces Retirement Home [AFRH] is an independent agency that was created in 1991 by the merging of the United States Soldiers' and

Airmen's Home in Washington D.C. and the United States Naval Home in Gulfport, Mississippi. AFRH operates and maintains the two communities to provide affordable resident services and care to retired and former enlisted servicemembers and their spouses.

**Trust Fund**

| | |
|---|---|
| Appropriations, 2024 | $77,000,000 |
| Budget estimate, 2025 | 69,520,000 |
| Committee recommendation | 69,520,000 |

**Committee Recommendation**

The Committee recommends authority to expend $69,520,000 from the Armed Forces Retirement Home Trust Fund to operate and maintain the Armed Forces Retirement Home-Washington, DC, and the Armed Forces Retirement Home-Gulfport, Mississippi. This amount is equal to the budget request.

The Committee recognizes the critical role of AFRH operations and the importance of delivering uninterrupted healthcare to its residents. Therefore, the Committee continues to support the 2- year availability of funds.

**Construction**

| | |
|---|---|
| Appropriations, 2024. | |
| Budget estimate, 2025 | 31,000,000 |
| Committee recommendation | 31,000,000 |

**Committee Recommendation**

The Committee recommends $31,000,000 for the Construction account. This is equal to the budget request and provides additional funding for the renovation of the Sheridan Building, which houses 80 percent of the residents at the Armed Forces Retirement Home-Washington campus, and was originally constructed in the 1960s.

## Administrative Provision

SEC. 301. The Committee includes a provision making available funds as authorized by 10 U.S.C. 7727.

## Title IV

*Federal Trust and Treaty Responsibilities.*—The Committee reminds agencies funded by this act of their obligation to uphold the Federal trust and treaty responsibilities to Tribes and Federal obligations to the Native Hawaiian Community. This includes upholding treaty and reserved rights, and any other rights and obligations under Federal law; supporting self-determination efforts by Native communities; fulfilling obligations under Presidential Memoranda and Executive Orders; and conducting early and robust government-to-government consultation with Tribes, and meaningful outreach and engagement with Native Hawaiians.

## General Provisions

SEC. 401. The Committee includes a provision that prohibits the obligation of funds beyond the current fiscal year unless expressly so provided.

SEC. 402. The Committee includes a provision that prohibits the use of funds for programs, projects, or activities not in compliance with Federal law relating to risk assessment, the protection of private property rights, or unfunded mandates.

SEC. 403. The Committee includes a provision that encourages the expansion of E-commerce technologies and procedures.

SEC. 404. The Committee includes a provision that specifies the congressional committees that are to receive all reports and notifications.

SEC. 405. The Committee includes a provision that limits funds from being transferred from this appropriations measure to any instrumentality of the United States Government without authority from an appropriations act.

SEC. 406. The Committee includes a provision regarding the posting of congressional reports on agency Web sites.

SEC. 407. The Committee includes a provision prohibiting the use of funds to establish or maintain a computer network unless such network blocks the viewing, downloading, and exchanging of pornography, except for law enforcement investigation, prosecution, or adjudication activities.

SEC. 408. The Committee includes a provision prohibiting the use of funds for the payment of first-class travel by an employee of the executive branch.

SEC. 409. The Committee includes a provision prohibiting the use of funds in this act for any contract where the contractor has not complied with E–Verify requirements.

SEC. 410. The Committee includes a provision prohibiting the use of funds in this act to construct facilities on military installations that do not meet resiliency standards.

SEC. 411. The Committee includes a provision prohibiting the use of funds in this act for the renovation, expansion, or construction of any facility in the continental United States for the purpose of housing any individual who has been detained at the United States Naval Station, Guantanamo Bay, Cuba.

## Program, Project, and Activity

In fiscal year 2021, for purposes of the Balanced Budget and Emergency Deficit Control Act of 1985 (Public Law 99–177) or the Balanced Budget and Emergency Deficit Control Reaffirmation Act of 1987 (Public Law 100–119), the following information provides the definition of the term ''program, project, and activity'' for departments, agencies and programs under the jurisdiction of the Military Construction and Veterans Affairs, and Related Agencies subcommittee. The term ''program, project, and activity'' shall include the most specific level of budget items identified in the Military Construction, Veterans Affairs, and Related Agencies Appropriations Act, 2021, the House and Senate Committee reports, and the conference report and accompanying joint explanatory statement of managers of the committee of conference.

If a sequestration order is necessary, in implementing the Presidential order, departments, and agencies shall apply any percentage reduction required for fiscal year 2021 pursuant to the provisions of Public Law 99–177 or Public Law 100–119 to all items specified in the justifications submitted to the Committees on Appropriations of the Senate and House of Representatives in support of the fiscal year 2021 budget estimates, as amended, for such departments and agencies, as modified by congressional action, and in addition, for the Department of Defense, Military Construction, the definition shall include specific construction locations as identified in the explanatory notes.

## Compliance with Paragraph 7, Rule XVI, of the Standing Rules of the Senate

Paragraph 7 of rule XVI requires that Committee reports accompanying general appropriations bills identify each recommended amendment which proposes an item of appropriation which is not made to carry out the provisions of an existing law, a treaty stipulation, or an act or resolution previously passed by the Senate during that session.

The Committee recommends funding for the following programs which currently lack authorization:

### Title I: Department of Defense

Military Construction, Army
Military Construction, Navy and Marine Corps
Military Construction, Air Force
Military Construction, Defense-Wide
Military Construction, Army National Guard
Military Construction, Air National Guard
Military Construction, Army Reserve
Military Construction, Navy Reserve
Military Construction, Air Force Reserve
North Atlantic Treaty Organization Security Investment Program
Department of Defense Base Closure Account
Family Housing Operation and Maintenance, Army
Family Housing Operation and Maintenance, Navy and Marine Corps
Family Housing Operation and Maintenance, Air Force
Family Housing Operation and Maintenance, Defense-Wide
Family Housing Construction, Army

Family Housing Construction, Navy and Marine Corps
Family Housing Construction, Air Force
Department of Defense Family Housing Improvement Fund
Department of Defense Military Unaccompanied Housing Improvement Fund

**Title II: Department of Veterans Affairs**

Veterans Benefits Administration
Veterans Health Administration
National Cemetery Administration
Departmental Administration

**Title III: Related Agencies**

American Battle Monuments Commission
U.S. Court of Appeals for Veterans Claims
Cemeterial Expenses, Army
Armed Forces Retirement Home

## Compliance with Paragraph 7(C), Rule XXVI of the Standing Rules of the Senate

Pursuant to paragraph 7(c) of rule XXVI, on July 11, 2024, the Committee ordered favorably reported a bill (S. 4677) making appropriations for military construction, the Department of Veterans Affairs, and related agencies for the fiscal year ending September 30, 2025, and for other purposes, provided, that the bill be subject to amendment and that any amendment increasing budget authority be offset by a reduction of equal or greater budget authority, by a recorded vote of 27–0, a quorum being present. The vote was as follows:

| Yeas | Nays |
|---|---|
| Chair Murray | |
| Mr. Durbin | |
| Mr. Reed | |

Mr. Tester
Mrs. Shaheen
Mr. Merkley
Mr. Coons
Mr. Schatz
Ms. Baldwin
Mr. Murphy
Mr. Manchin
Mr. Van Hollen
Mr. Heinrich
Mr. Peters
Ms. Sinema
Ms. Collins
Mr. McConnell
Ms. Murkowski
Mr. Graham
Mr. Moran
Mr. Hoeven
Mr. Boozman
Mr. Kennedy
Mrs. Hyde-Smith
Mr. Hagerty
Mrs. Britt
Mrs. Fischer

## Compliance with Paragraph 12, Rule XXVI of the Standing Rules of the Senate

Paragraph 12 of rule XXVI requires that Committee reports on a bill or joint resolution repealing or amending any statute or part of any statute include ''(a) the text of the statute or part thereof which is proposed to be repealed; and (b) a comparative print of that part of the bill or joint resolution making the amendment and of the statute or part thereof proposed to be amended, showing by stricken-through type and italics, parallel columns, or other appropriate typographical devices the omissions and insertions which would be made by the bill or joint resolution if enacted in the form recommended by the committee.''

The Committee bill as recommended contains no such provisions.

## Budgetary Impact of Bill

### Prepared in Consultation with the Congressional Budget Office Pursuant to Sec. 308(A), Public Law 93–344, as Amended

[In millions of dollars]

| | Budget authority | | Outlays | |
|---|---|---|---|---|
| | Committee allocation | Amount in bill | Committee allocation | Amount in bill |
| Comparison of amounts in the bill with the subcommittee allocation for 2025: Military Construction, Veterans Affairs, and Related Agencies: | | | | |
| Mandatory | 221,911 | 221,911 | 218,068 | [1] 218,068 |
| Discretionary | 148,876 | 148,876 | 144,176 | [1] 144,166 |
| Defense | 19,307 | 19,307 | | |
| Non-defense | 129,569 | 129,569 | | |
| Projection of outlays associated with the recommendation: | | | | |
| 2025 | | | | [2] 212,753 |
| 2026 | | | | 124,290 |
| 2027 | | | | 18,753 |
| 2028 | | | | 6,140 |
| 2029 and future years | | | | 5,896 |
| Financial assistance to State and local governments for 2025 | NA | 290 | NA | [2] 271 |

1 Includes outlays from prior-year budget authority.
2 Excludes outlays from prior-year budget authority.
NA: Not applicable.

## Military Construction Project Listing by Location

[In thousands of dollars]

| Installation and project | Budget estimate | Committee recommendation | Committee recommendation compared to budget estimate |
|---|---|---|---|
| ALABAMA | | | |
| ARMY: | | | |
| FORT NOVOSEL: | | | |
| CRASH/RESCUE STATION AT HATCH ARMY HELIPORT: MINOR CONSTRUCTION | .................... | 6,500 | +6,500 |
| CRASH/RESCUE STATION AT TABERNACLE STAGEFIELD: MINOR CONSTRUCTION | .................... | 6,600 | +6,600 |
| FLAT IRON SUPPORT FACILITY: MINOR CONSTRUCTION | .................... | 5,500 | +5,500 |
| FLIGHT CONTROL TOWER—HANCHEY ARMY AIRFIELD: DESIGN | .................... | 1,300 | +1,300 |
| INSTALLATION SERVICE NODE FACILITY: DESIGN | .................... | 2,600 | +2,600 |
| DEFENSE-WIDE: | | | |
| REDSTONE ARSENAL: | | | |
| GROUND TEST FACILITY INFRASTRUCTURE (INC) | 80,000 | .................... | −80,000 |
| TOTAL, ALABAMA | 80,000 | 22,500 | −57,500 |
| ALASKA | | | |
| ARMY: | | | |
| FORT WAINWRIGHT: | | | |
| AUTOMATED MULTIPURPOSE MACHINE GUN RANGE | 23,000 | 23,000 | .................... |
| DINING FACILITY: DESIGN | .................... | 7,700 | +7,700 |
| TACTICAL EQUIPMENT MAINTENANCE FACILITY: DESIGN | .................... | 10,800 | +10,800 |
| AIR FORCE: | | | |
| EIELSON AIR FORCE BASE: | | | |
| JOINT MOBILITY CENTER EXPANSION: DESIGN | .................... | 5,800 | +5,800 |
| PERMANENT PARTY DORMITORY: DESIGN | .................... | 6,700 | +6,700 |
| JOINT BASE ELMENDORF-RICHARDSON | | | |
| JOINT INTEGRATED TEST AND TRAINING CTR, INC | 126,000 | 126,000 | .................... |
| PRECISION GUIDED MUNITIONS COMPLEX: DESIGN | .................... | 15,600 | +15,600 |
| DEFENSE-WIDE: | | | |
| EIELSON AIR FORCE BASE: | | | |
| FUEL OPERATIONS AND LAB FACILITY | 14,000 | 14,000 | .................... |
| JOINT BASE ELMENDORF-RICHARDSON: | | | |
| FUEL FACILITIES | 55,000 | 55,000 | .................... |
| ARMY NATIONAL GUARD: | | | |
| FORT RICHARDSON: | | | |
| NATIONAL GUARD READINESS CENTER | 67,000 | 67,000 | .................... |
| AIR NATIONAL GUARD: | | | |
| JOINT BASE ELMENDORF-RICHARDSON: | | | |
| COMBAT RESCUE HELICOPTER SIMULATOR | 19,300 | 19,300 | .................... |
| TOTAL, ALASKA | 304,300 | 350,900 | +46,600 |
| ARIZONA | | | |
| ARMY: | | | |
| YUMA PROVING GROUND: | | | |
| RADAR OPERATIONS & MAINTENANCE FACILITY: MINOR CONSTRUCTION | .................... | 6,000 | +6,000 |
| NAVY AND MARINE CORPS: | | | |
| MARINE CORPS AIR STATION YUMA: | | | |
| UDP TRANSIENT BARRACKS: DESIGN | .................... | 3,800 | +3,800 |
| WATER TREATMENT PLANT | .................... | 90,000 | +90,000 |
| DEFENSE-WIDE: | | | |
| YUMA: | | | |
| SOF MILITARY FREE FALL ADVANCED TRAIN COMPLEX | 62,000 | 62,000 | .................... |

## Military Construction Project Listing by Location (Continued)

| Installation and project | Budget estimate | Committee recommendation | Committee recommendation compared to budget estimate |
|---|---|---|---|
| AIR NATIONAL GUARD: | | | |
| MORRIS AIR NATIONAL GUARD BASE: | | | |
| BASE ENTRY COMPLEX: COST TO COMPLETE | ............ | 7,000 | +7,000 |
| TOTAL, ARIZONA | 62,000 | 168,800 | +106,800 |
| CALIFORNIA | | | |
| ARMY: | | | |
| CONCORD: | | | |
| AMMUNITION HOLDING FACILITY | 68,000 | 68,000 | ............ |
| FORT IRWIN: | | | |
| TRAINING SUPPORT CENTER | 44,000 | 44,000 | ............ |
| AIR FORCE: | | | |
| BEALE AIR FORCE BASE: | | | |
| MULTI-DOMAIN OPERATIONS COMPLEX | ............ | 44,000 | +44,000 |
| VANDENBERG SPACE FORCE BASE: | | | |
| GBSD RE-ENTRY VEHICLE FACILITY | 110,000 | 110,000 | ............ |
| SENTINEL AETC FORMAL TRAINING UNIT | 167,000 | 167,000 | ............ |
| DEFENSE-WIDE: | | | |
| BRIDGEPORT: | | | |
| FUEL FACILITIES | 19,300 | 19,300 | ............ |
| CAMP PENDLETON: | | | |
| AMBULATORY CARE CENTER ADD/ALT (AREA 53) | 26,440 | 26,440 | ............ |
| AMBULATORY CARE CENTER ADD/ALT (AREA 62) | 24,930 | 24,930 | ............ |
| AMBULATORY CARE CENTER REPLACEMENT (AREA 22) | 45,040 | 45,040 | ............ |
| CORONADO: | | | |
| SOF OPERATIONS SUPPORT FACILITY PH 2 | 51,000 | ............ | −51,000 |
| AIR NATIONAL GUARD: | | | |
| MOFFETT FIELD: | | | |
| COMBAT RESCUE HELICOPTER SIMULATOR | 12,600 | 12,600 | ............ |
| ARMY RESERVE: | | | |
| BELL: | | | |
| ARMY RESERVE CENTER: DESIGN | ............ | 3,750 | +3,750 |
| CAMP PARKS: | | | |
| ADVANCED SKILLS TRAINING BARRACKS | 42,000 | 42,000 | ............ |
| TOTAL, CALIFORNIA | 610,310 | 607,060 | −3,250 |
| COLORADO | | | |
| AIR FORCE: | | | |
| BUCKLEY SPACE FORCE BASE: | | | |
| POWER INDEPENDENCE | ............ | 60,000 | +60,000 |
| DEFENSE-WIDE: | | | |
| FORT CARSON: | | | |
| AMBULATORY CARE CENTER REPLACEMENT | 41,000 | 41,000 | ............ |
| TOTAL, COLORADO | 41,000 | 101,000 | +60,000 |
| CONNECTICUT | | | |
| NAVY AND MARINE CORPS: | | | |
| NAVAL SUBMARINE BASE NEW LONDON: | | | |
| SUBMARINE STORAGE, MAINTENANCE & OPERATIONS FACILITY: DESIGN | ............ | 3,200 | +3,200 |
| ARMY NATIONAL GUARD: | | | |
| GROTON: | | | |
| AIRCRAFT MAINTENANCE HANGAR: DESIGN | ............ | 6,500 | +6,500 |

| Installation and project | Budget estimate | Committee recommendation | Committee recommendation compared to budget estimate |
|---|---|---|---|
| TOTAL, CONNECTICUT | ............ | 9,700 | +9,700 |
| DELAWARE | | | |
| ARMY NATIONAL GUARD: | | | |
| DAGSBORO: | | | |
| DAGSBORO READINESS CENTER: MINOR CONSTRUCTION | ............ | 7,500 | +7,500 |
| AIR NATIONAL GUARD: | | | |
| NEW CASTLE COUNTY AIRPORT: | | | |
| ENTRY CONTROL FACILITY: MINOR CONSTRUCTION | ............ | 8,000 | +8,000 |
| TOTAL, DELAWARE | ............ | 15,500 | +15,500 |
| FLORIDA | | | |
| NAVY AND MARINE CORPS: | | | |
| CAPE CANAVERAL SPACE FORCE STATION | | | |
| ENGINEERING TEST FACILITY | 221,060 | 221,060 | ............ |
| AIR FORCE: | | | |
| EGLIN AIR FORCE BASE: | | | |
| LRSO HARDWARE SOFTWARE DEVELOPMENT TEST FAC | 8,400 | 8,400 | ............ |
| DEFENSE-WIDE: | | | |
| HURLBURT FIELD: | | | |
| SOF AFSOC OPERATIONS FACILITY | 14,000 | 14,000 | ............ |
| AIR NATIONAL GUARD: | | | |
| JACKSONVILLE INTERNATIONAL AIRPORT: | | | |
| F–35 CONSOLIDATED WEAPONS TRAINING | 26,200 | 26,200 | ............ |
| TOTAL, FLORIDA | 269,660 | 269,660 | ............ |
| GEORGIA | | | |
| ARMY: | | | |
| FORT STEWART: | | | |
| ENLISTED UNACCOMPANIED PERSONNEL BARRACKS (PN 96725): DESIGN | ............ | 10,500 | +10,500 |
| FORT STEWART/HUNTER ARMY AIRFIELD: | | | |
| SOF MILITARY WORKING DOG KENNEL FACILITY: DESIGN | ............ | 1,230 | +1,230 |
| NAVY AND MARINE CORPS: | | | |
| KINGS BAY: | | | |
| TRIDENT REFIT FAC EXPAN (COLUM SUB—INC) | 115,000 | 115,000 | ............ |
| AIR FORCE: | | | |
| ROBINS AIR FORCE BASE: | | | |
| BATTLE MANAGEMENT COMBINED OPS COMPLEX, INC | 64,000 | 57,500 | −6,500 |
| DEFENSE-WIDE: | | | |
| HUNTER ARMY AIRFIELD: | | | |
| SOF CONSOLIDATED RIGGING FACILITY | 47,000 | 47,000 | ............ |
| SOF MILITARY WORKING DOG KENNEL FACILITY | 16,800 | 16,800 | ............ |
| ARMY RESERVE: | | | |
| DOBBINS AIR RESERVE BASE: | | | |
| ARMY RESERVE CENTER | 78,000 | 78,000 | ............ |
| MARINE CORPS LOGISTICS BASE ALBANY: | | | |
| ARMY RESERVE CENTER | ............ | 40,000 | +40,000 |
| AIR FORCE RESERVE: | | | |
| DOBBINS AIR RESERVE BASE: | | | |
| SECURITY FORCES FACILITY | 22,000 | ............ | −22,000 |
| TOTAL, GEORGIA | 342,800 | 366,030 | +23,230 |

## Military Construction Project Listing by Location (Continued)

| Installation and project | Budget estimate | Committee recommendation | Committee recommendation compared to budget estimate |
|---|---|---|---|
| HAWAII | | | |
| ARMY: | | | |
| POHAKULOA TRAINING AREA: | | | |
| AIRFIELD OPERATIONS BUILDING | .................. | 20,000 | +20,000 |
| WHEELER ARMY AIR FIELD: | | | |
| AIRCRAFT MAINTENANCE HANGAR | 231,000 | 231,000 | .................. |
| NAVY AND MARINE CORPS: | | | |
| JOINT BASE PEARL HARBOR–HICKAM: | | | |
| DRY DOCK 3 REPLACEMENT (INC) | 1,199,000 | 1,199,000 | .................. |
| (RED HILL) WATER TREATMENT PLANT | .................. | 90,000 | +90,000 |
| KANEOHE BAY: | | | |
| AIRCRAFT HANGAR & PARKING APRON | 203,520 | 203,520 | .................. |
| AIR FORCE: | | | |
| JOINT BASE PEARL HARBOR–HICKAM: | | | |
| GENERAL PURPOSE MAINTENANCE HANGAR: DESIGN | .................. | 7,000 | +7,000 |
| MOBILITY PROCESSING CENTER/WAR RESERVE MATERIAL WAREHOUSE: DESIGN | .................. | 11,000 | +11,000 |
| AIR NATIONAL GUARD: | | | |
| HICKAM AIR FORCE BASE: | | | |
| SPACE CONTROL CENTER | 36,600 | 36,600 | .................. |
| TOTAL, HAWAII | 1,670,120 | 1,798,120 | +128,000 |
| IDAHO | | | |
| AIR FORCE: | | | |
| MOUNTAIN HOME AIR FORCE BASE: | | | |
| CHILD DEVELOPMENT CENTER | 40,000 | 40,000 | .................. |
| TOTAL, IDAHO | 40,000 | 40,000 | .................. |
| ILLINOIS | | | |
| ARMY: | | | |
| ROCK ISLAND ARSENAL: | | | |
| FORGING EQUIPMENT ANNEX: MINOR CONSTRUCTION | .................. | 5,000 | +5,000 |
| NEW WEAPONS QUALITY ASSURANCE/CALIBRATION FACILITY: MINOR CONSTRUCTION | .................. | 4,250 | +4,250 |
| ARMY NATIONAL GUARD: | | | |
| MARSEILLES TRAINING CENTER: | | | |
| MARSEILLES TRAINING CENTER RANGE RENOVATION: MINOR CONSTRUCTION | .................. | 4,300 | +4,300 |
| TOTAL, ILLINOIS | .................. | 13,550 | +13,550 |
| INDIANA | | | |
| AIR FORCE RESERVE: | | | |
| GRISSOM AIR RESERVE BASE: | | | |
| INDOOR SMALL ARMS RANGE | 21,000 | 21,000 | .................. |
| TOTAL, INDIANA | 21,000 | 21,000 | .................. |
| IOWA | | | |
| ARMY NATIONAL GUARD: | | | |
| SIOUX CITY ARMORY: | | | |
| NATIONAL GUARD VEHICLE MAINTENANCE SHOP | 13,800 | 13,800 | .................. |

| Installation and project | Budget estimate | Committee recommendation | Committee recommendation compared to budget estimate |
|---|---|---|---|
| TOTAL, IOWA | 13,800 | 13,800 | ............ |
| KANSAS | | | |
| ARMY: | | | |
| FORT LEAVENWORTH: | | | |
| B95 FIRE STATION TRUCK BAY ADDITION: MINOR CONSTRUCTION | ............ | 8,200 | +8,200 |
| FIRE STATION HEADQUARTERS: DESIGN | ............ | 1,800 | +1,800 |
| FUNCTIONAL FITNESS CENTER: DESIGN | ............ | 1,700 | +1,700 |
| TOTAL, KANSAS | ............ | 11,700 | +11,700 |
| KENTUCKY | | | |
| ARMY: | | | |
| BLUE GRASS ARMY DEPOT: | | | |
| MILITARY CONTAINER PRODUCTION FACILITY: DESIGN | ............ | 27,000 | +27,000 |
| FORT CAMPBELL: | | | |
| AUTOMATED RECORD FIRE PLUS RANGE | 11,800 | 11,800 | ............ |
| FOUNDRY TRAINING FACILITY: MINOR CONSTRUCTION | ............ | 8,100 | +8,100 |
| MODERNIZED HANGAR: DESIGN | ............ | 11,000 | +11,000 |
| AIR NATIONAL GUARD: | | | |
| LOUISVILLE INTERNATIONAL AIRPORT–STANDIFORD FLD: | | | |
| C–130J ADAL FUEL CELL HANGAR: MINOR CONSTRUCTION | ............ | 8,500 | +8,500 |
| ARMY RESERVE: | | | |
| FORT KNOX: | | | |
| AVIATION SUPPORT FACILITY | ............ | 75,000 | +75,000 |
| TOTAL, KENTUCKY | 11,800 | 141,400 | +129,600 |
| LOUISIANA | | | |
| ARMY: | | | |
| FORT JOHNSON: | | | |
| BARRACKS | 117,000 | ............ | −117,000 |
| ROTATIONAL UNIT BILLETING AREA: DESIGN | ............ | 9,900 | +9,900 |
| AIR FORCE: | | | |
| BARKSDALE AIR FORCE BASE: | | | |
| CHILD DEVELOPMENT CENTER | ............ | 22,000 | +22,000 |
| ARMY NATIONAL GUARD: | | | |
| LAFAYETTE READINESS CENTER: | | | |
| NATIONAL GUARD READINESS CENTER | 33,000 | 33,000 | ............ |
| TOTAL, LOUISIANA | 150,000 | 64,900 | −85,100 |
| MAINE | | | |
| NAVY AND MARINE CORPS: | | | |
| PORTSMOUTH NAVAL SHIPYARD: | | | |
| MULTI-MISSION DRYDOCK #1 EXTENSION (INC) | 400,578 | 400,578 | ............ |
| PARKING STRUCTURE: DESIGN | ............ | 9,500 | +9,500 |
| ARMY NATIONAL GUARD: | | | |
| SACO: | | | |
| SOUTHERN MAINE READINESS CENTER: DESIGN | ............ | 800 | +800 |
| WOODVILLE: | | | |
| AUTOMATED QUALIFICATION TRAINING RANGE: DESIGN | ............ | 1,000 | +1,000 |

## Military Construction Project Listing by Location (Continued)

| Installation and project | Budget estimate | Committee recommendation | Committee recommendation compared to budget estimate |
|---|---|---|---|
| AIR NATIONAL GUARD: | | | |
| BANGOR INTERNATIONAL AIRPORT: | | | |
| AIRCRAFT MAINTENANCE SQUADRON/AIRCRAFT GROUND EQUIPMENT FACILITY: DESIGN | .................... | 800 | + 800 |
| FUEL CELL HANGAR | .................... | 50,000 | + 50,000 |
| TOTAL, MAINE | 400,578 | 462,678 | + 62,100 |
| MARYLAND | | | |
| ARMY: | | | |
| FORT MEADE: | | | |
| CHILD DEVELOPMENT CENTER | 46,000 | 46,000 | .................... |
| NAVY AND MARINE CORPS: | | | |
| NAVAL SURFACE WARFARE CENTER INDIAN HEAD: | | | |
| CONTAINED BURN FACILITY | .................... | 50,000 | + 50,000 |
| DEFENSE-WIDE: | | | |
| BETHESDA NAVAL HOSPITAL: | | | |
| MEDCEN ADDITION/ALTERATION INCR 8 | 77,651 | 77,651 | .................... |
| FORT MEADE: | | | |
| NSAW EAST CAMPUS BUILDING #5, INC 2 | 265,000 | 265,000 | .................... |
| JOINT BASE ANDREWS: | | | |
| AMBULATORY CARE CENTER (INC) | 15,040 | 15,040 | .................... |
| ARMY NATIONAL GUARD: | | | |
| HAGERSTOWN: | | | |
| FIELD MAINTENANCE SHOP: DESIGN | .................... | 2,470 | + 2,470 |
| TOTAL, MARYLAND | 403,691 | 456,161 | + 52,470 |
| MASSACHUSETTS | | | |
| AIR FORCE: | | | |
| HANSCOM AIR FORCE BASE: | | | |
| MIT–LL/ENGINEERING AND PROTOTYPE FAC, INC | 76,000 | 76,000 | .................... |
| TOTAL, MASSACHUSETTS | 76,000 | 76,000 | .................... |
| MICHIGAN | | | |
| ARMY: | | | |
| DETROIT ARSENAL: | | | |
| MANNED/UNMANNED TACTICAL VEHICLE LAB | 37,000 | 37,000 | .................... |
| ARMY NATIONAL GUARD: | | | |
| OLYMPIA ARMORY: | | | |
| READINESS CENTER ADDITION/ALTERATION: DESIGN | .................... | 4,400 | + 4,400 |
| TOTAL, MICHIGAN | 37,000 | 41,400 | + 4,400 |
| MINNESOTA | | | |
| AIR NATIONAL GUARD: | | | |
| MINNEAPOLIS ST PAUL INTERNATIONAL AIRPORT: | | | |
| C–130J FUEL CELL MAINTENANCE HANGAR: DESIGN | .................... | 4,300 | + 4,300 |
| TOTAL, MINNESOTA | .................... | 4,300 | + 4,300 |

| Installation and project | Budget estimate | Committee recommendation | Committee recommendation compared to budget estimate |
|---|---|---|---|
| MISSISSIPPI | | | |
| ARMY: | | | |
| ARMY ENGINEER RESEARCH AND DEVELOPMENT CENTER: | | | |
| JET ENGINE THERMAL SIMULATOR: DESIGN | ............ | 700 | +700 |
| AIR FORCE: | | | |
| KEESLER AIR FORCE BASE: | | | |
| AIR TRAFFIC CONTROL TOWER | ............ | 25,000 | +25,000 |
| DEFENSE-WIDE: | | | |
| KEY FIELD: | | | |
| UPGRADE FUEL HYDRANT SYSTEM: DESIGN | ............ | 1,000 | +1,000 |
| ARMY NATIONAL GUARD: | | | |
| SOUTHAVEN READINESS CENTER: | | | |
| NATIONAL GUARD READINESS CENTER | 33,000 | 11,000 | −22,000 |
| AIR NATIONAL GUARD: | | | |
| KEY FIELD: | | | |
| FUEL CELL/CORROSION CONTROL HANGAR: DESIGN | ............ | 5,000 | +5,000 |
| MAINTENANCE HANGAR: DESIGN | ............ | 5,000 | +5,000 |
| TOTAL, MISSISSIPPI | 33,000 | 47,700 | +14,700 |
| MISSOURI | | | |
| ARMY: | | | |
| FORT LEONARD WOOD: | | | |
| ADV INDIVIDUAL TRAINING BARRACKS CPLX, PH2 | 144,000 | 144,000 | ............ |
| DEFENSE-WIDE: | | | |
| WHITEMAN AIR FORCE BASE: | | | |
| FLIGHTLINE FUELING FACILITIES | 19,500 | 19,500 | ............ |
| TOTAL, MISSOURI | 163,500 | 163,500 | ............ |
| MONTANA | | | |
| AIR FORCE: | | | |
| MALMSTROM AIR FORCE BASE: | | | |
| GBSD COMMERCIAL ENTRANCE CONTROL FACILITY | 20,000 | 20,000 | ............ |
| WEAPONS STORAGE & MAINTENANCE FAC INC | 238,000 | 238,000 | ............ |
| ARMY NATIONAL GUARD: | | | |
| MALTA READINESS CENTER: | | | |
| NATIONAL GUARD VEHICLE MAINTENANCE SHOP | 14,800 | 14,800 | ............ |
| TOTAL, MONTANA | 272,800 | 272,800 | ............ |
| NEBRASKA | | | |
| AIR FORCE: | | | |
| OFFUTT AIR FORCE BASE: | | | |
| CONSOLIDATED TRAINING COMPLEX/PROFESSIONAL DEVELOPMENT CENTER: DESIGN | ............ | 6,000 | +6,000 |
| SURVIVABLE AIRBORNE OPERATIONS CENTER COMPLEX: DESIGN | ............ | 70,000 | +70,000 |
| ARMY NATIONAL GUARD: | | | |
| HASTINGS: | | | |
| COLLECTIVE TRAINING UNACCOMPANIED HOUSING: MINOR CONSTRUCTION | ............ | 5,900 | +5,900 |
| TOTAL, NEBRASKA | ............ | 81,900 | +81,900 |

# Military Construction Project Listing by Location (Continued)

| Installation and project | Budget estimate | Committee recommendation | Committee recommendation compared to budget estimate |
|---|---|---|---|
| NEVADA | | | |
| NAVY AND MARINE CORPS: | | | |
| FALLON: | | | |
| TRAINING RANGE LAND ACQUISITION—PHASE 2 | 48,300 | 48,300 | |
| AIR FORCE: | | | |
| NELLIS AIR FORCE BASE: | | | |
| REGIONAL COMMUNICATIONS CENTER: DESIGN | | 3,500 | +3,500 |
| ARMY NATIONAL GUARD: | | | |
| HAWTHORNE ARMY DEPOT: | | | |
| AUTOMATED QUALIFICATION/TRAINING RANGE | 18,000 | 18,000 | |
| AIR NATIONAL GUARD | | | |
| RENO–TAHOE INTERNATIONAL AIRPORT: | | | |
| AIRCRAFT PARKING RAMP EXTENSION: DESIGN | | 1,000 | +1,000 |
| AIRCRAFT SUPPORT EQUIPMENT AND ENGINE FACILITY: DESIGN | | 2,000 | +2,000 |
| FUEL CELL HANGAR: DESIGN | | 3,000 | +3,000 |
| TOTAL, NEVADA | 66,300 | 75,800 | +9,500 |
| NEW HAMPSHIRE | | | |
| ARMY NATIONAL GUARD: | | | |
| CONCORD: | | | |
| LAKOTA HANGAR ADDITION AND ENVIRONMENTAL UPGRADES: DESIGN | | 1,200 | +1,200 |
| PEMBROOKE: | | | |
| READINESS CENTER ADDITION/ALTERATION: DESIGN | | 6,500 | +6,500 |
| TOTAL, NEW HAMPSHIRE | | 7,700 | +7,700 |
| NEW JERSEY | | | |
| ARMY NATIONAL GUARD: | | | |
| VINELAND: | | | |
| NATIONAL GUARD VEHICLE MAINTENANCE SHOP | 23,000 | 23,000 | |
| AIR NATIONAL GUARD: | | | |
| ATLANTIC CITY INTERNATIONAL AIRPORT: | | | |
| F–16 MISSION TRAINING CENTER | 18,000 | 18,000 | |
| FIGHTER OPERATIONS FACILITY: DESIGN | | 3,400 | +3,400 |
| ARMY RESERVE: | | | |
| JOINT BASE MCGUIRE–DIX–LAKEHURST | | | |
| VERTICAL SKILLS FACILITY | 16,000 | 16,000 | |
| TOTAL, NEW JERSEY | 57,000 | 60,400 | +3,400 |
| NEW MEXICO | | | |
| ARMY: | | | |
| WHITE SANDS MISSILE RANGE: | | | |
| NORTH RANGE MISSION CONTROL CENTER: DESIGN | | 3,330 | +3,330 |
| AIR FORCE: | | | |
| CANNON AIR FORCE BASE: | | | |
| ADAL SECURITY FORCES FACILITY: MINOR CONSTRUCTION | | 4,000 | +4,000 |
| SPECIAL OPERATIONS FORCES DAGRE FACILITY: MINOR CONSTRUCTION | | 7,200 | +7,200 |
| HOLLOMAN AIR FORCE BASE: | | | |
| HOLLOMAN HIGH–SPEED TEST TRACK: DESIGN | | 22,400 | +22,400 |

| Installation and project | Budget estimate | Committee recommendation | Committee recommendation compared to budget estimate |
|---|---|---|---|
| TOTAL, NEW MEXICO | .................... | 36,930 | + 36,930 |
| NEW YORK | | | |
| ARMY: | | | |
| FORT DRUM: | | | |
| AIRCRAFT MAINTENANCE HANGAR ADDITION: DESIGN | .................... | 9,300 | + 9,300 |
| FIELD ARTILLERY VEHICLE STORAGE SHEDS: DESIGN | .................... | 800 | + 800 |
| FIRE STATION 3 (WSAAF): DESIGN | .................... | 2,760 | + 2,760 |
| ORTC TRAINING BARRACKS: DESIGN | .................... | 7,790 | + 7,790 |
| WATERVLIET ARSENAL: | | | |
| FIRE STATION | 53,000 | 53,000 | .................... |
| AIR NATIONAL GUARD: | | | |
| FRANCIS S GABRESKI AIRPORT: | | | |
| COMBAT RESCUE HELICOPTER SIMULATOR | 14,000 | 14,000 | .................... |
| AIR FORCE RESERVE: | | | |
| NIAGARA FALLS AIR RESERVE STATION: | | | |
| TWY A4 AND RUNWAY 28R OVERRUN: DESIGN | .................... | 5,400 | + 5,400 |
| TOTAL, NEW YORK | 67,000 | 93,050 | + 26,050 |
| NORTH CAROLINA | | | |
| ARMY: | | | |
| FORT LIBERTY: | | | |
| CHILD DEVELOPMENT CENTER | 39,000 | 3,000 | − 36,000 |
| NAVY AND MARINE CORPS: | | | |
| CHERRY POINT MARINE CORPS AIR STATION: | | | |
| F–35 AIRCRAFT SUSTAINMENT CTR, INC | 50,000 | 50,000 | .................... |
| AIRCRAFT MAINTENANCE HANGAR | 213,520 | 213,520 | .................... |
| COMPOSITE REPAIR FACILITY | 114,020 | 114,020 | .................... |
| AIR FORCE: | | | |
| SEYMOUR JOHNSON AIR FORCE BASE: | | | |
| COMBAT ARMS TRAINING AND MAINTENANCE COMPLEX | .................... | 15,000 | + 15,000 |
| DEFENSE-WIDE: | | | |
| CAMP LEJEUNE: | | | |
| SOF ARMORY | 25,400 | 25,400 | .................... |
| FORT LIBERTY: | | | |
| SOF ARMS ROOM ADDITION | 11,800 | 11,800 | .................... |
| TOTAL, NORTH CAROLINA | 453,740 | 432,740 | − 21,000 |
| OHIO | | | |
| AIR FORCE: | | | |
| WRIGHT–PATTERSON AIR FORCE BASE: | | | |
| ADVANCED MATERIALS RESEARCH LABORATORY—C2A: DESIGN | .................... | 9,200 | + 9,200 |
| DORMITORY: DESIGN | .................... | 3,400 | + 3,400 |
| US SPACE FORCE INTELLIGENCE CENTER: DESIGN | .................... | 19,000 | + 19,000 |
| AIR FORCE RESERVE: | | | |
| YOUNGSTOWN AIR RESERVE STATION: | | | |
| BASE FIRE STATION | 25,000 | 25,000 | .................... |
| TOTAL, OHIO | 25,000 | 56,600 | + 31,600 |

# Military Construction Project Listing by Location (Continued)

| Installation and project | Budget estimate | Committee recommendation | Committee recommendation compared to budget estimate |
|---|---|---|---|
| OKLAHOMA | | | |
| AIR FORCE: | | | |
| ALTUS AIR FORCE BASE: | | | |
| MAIN GATE ENTRY CONTROL FACILITY: DESIGN | .................... | 2,200 | +2,200 |
| ARMY NATIONAL GUARD: | | | |
| SHAWNEE READINESS CENTER: | | | |
| NATIONAL GUARD READINESS CENTER | 29,000 | 29,000 | .................... |
| TOTAL, OKLAHOMA | 29,000 | 31,200 | +2,200 |
| OREGON | | | |
| AIR FORCE: | | | |
| MOUNTAIN HOME AIR FORCE BASE: | | | |
| HOMELAND DEFENSE OVER-THE-HORIZON RADAR, INC | 198,000 | 198,000 | .................... |
| DEFENSE-WIDE: | | | |
| KLAMATH FALLS AIRPORT—KINGSLEY FIELD: | | | |
| FUEL FACILITY: DESIGN | .................... | 3,000 | +3,000 |
| ARMY NATIONAL GUARD: | | | |
| NAVAL WEAPONS SYSTEMS TRAINING FACILITY BOARDMAN: | | | |
| MULTI-PURPOSE MACHINE GUN RANGE: DESIGN | .................... | 1,690 | +1,690 |
| AIR NATIONAL GUARD: | | | |
| KLAMATH FALLS AIRPORT—KINGSLEY FIELD: | | | |
| AIRCRAFT SHELTER MCCA: DESIGN | .................... | 4,300 | +4,300 |
| TOTAL, OREGON | 198,000 | 206,990 | +8,990 |
| PENNSYLVANIA | | | |
| ARMY: | | | |
| LETTERKENNY ARMY DEPOT: | | | |
| COMPONENT REBUILD SHOP (INC 1) | 90,000 | 90,000 | .................... |
| MISSILE/MUNITIONS DISTRIBUTION FACILITY | 62,000 | 62,000 | .................... |
| ARMY NATIONAL GUARD: | | | |
| FORT INDIANTOWN GAP TRAINING SITE: | | | |
| READY BUILDING: DESIGN | .................... | 3,250 | +3,250 |
| AIR NATIONAL GUARD: | | | |
| PITTSBURGH INTERNATIONAL AIRPORT: | | | |
| ENTRY CONTROL FACILITY: MINOR CONSTRUCTION | .................... | 5,000 | +5,000 |
| ARMY RESERVE: | | | |
| NEW CASTLE: | | | |
| AREA MAINTENANCE SUPPORT ACTIVITY AND LAND: DESIGN | .................... | 1,750 | +1,750 |
| WILKES-BARRE: | | | |
| AREA MAINTENANCE SUPPORT ACTIVITY EQUIPMENT | 22,000 | 22,000 | .................... |
| TOTAL, PENNSYLVANIA | 174,000 | 184,000 | +10,000 |
| RHODE ISLAND | | | |
| ARMY NATIONAL GUARD: | | | |
| NORTH KINGSTOWN: | | | |
| NATIONAL GUARD READINESS CENTER: COST TO COMPLETE | .................... | 16,000 | +16,000 |
| TOTAL, RHODE ISLAND | .................... | 16,000 | +16,000 |
| SOUTH CAROLINA | | | |
| NAVY AND MARINE CORPS: | | | |
| MARINE CORPS AIR STATION BEAUFORT: | | | |
| AIRCRAFT MAINTENANCE HANGAR: COST TO COMPLETE | .................... | 61,700 | +61,700 |

| Installation and project | Budget estimate | Committee recommendation | Committee recommendation compared to budget estimate |
|---|---|---|---|
| DEFENSE-WIDE: | | | |
| MARINE CORPS AIR STATION BEAUFORT: | | | |
| FUEL PIER | 31,500 | 31,500 | ............ |
| PARRIS ISLAND: | | | |
| AMBULATORY CARE CLINIC REPLACEMENT (DENTAL) | 72,050 | 72,050 | ............ |
| AIR FORCE RESERVE: | | | |
| JOINT BASE CHARLESTON: | | | |
| AEROMEDICAL EVACUATION FACILITY | ............ | 33,000 | +33,000 |
| TOTAL, SOUTH CAROLINA | 103,550 | 198,250 | +94,700 |
| SOUTH DAKOTA | | | |
| AIR FORCE: | | | |
| ELLSWORTH AIR FORCE BASE: | | | |
| B–21 ADAL SQUADRON OPERATIONS | 44,000 | 44,000 | ............ |
| B–21 E ALERT APRON ENV PROTECTION SHELTERS | 79,000 | 79,000 | ............ |
| B–21 N ENV PROTECTION SHELTERS (60 ROW) | 54,000 | 54,000 | ............ |
| B–21 WEAPONS GENERATION FACILITY INC | 105,000 | 105,000 | ............ |
| TOTAL, SOUTH DAKOTA | 282,000 | 282,000 | ............ |
| TEXAS | | | |
| ARMY: | | | |
| FORT BLISS: | | | |
| RAIL YARD: COST TO COMPLETE | 44,000 | 44,000 | ............ |
| RED RIVER ARMY DEPOT: | | | |
| VEHICLE PAINT SHOP | 34,000 | 34,000 | ............ |
| AIR FORCE: | | | |
| DYESS AIR FORCE BASE: | | | |
| B–21 LRS FUELS ADMINISTRATIVE LABORATORY | 12,800 | 12,800 | ............ |
| B–21 REFUELER TRUCK YARD | 18,500 | 18,500 | ............ |
| JOINT BASE SAN ANTONIO: | | | |
| METC—BARRACKS/SHIPS/DORMS #1, INC | 77,000 | 77,000 | ............ |
| LAUGHLIN AIR FORCE BASE: | | | |
| T–7A GROUND BASED TRAINING SYSTEM FACILITY | 38,000 | 38,000 | ............ |
| T–7A UNIT MAINTENANCE TRAINING FACILITY | 18,000 | 18,000 | ............ |
| DEFENSE-WIDE: | | | |
| CORPUS CHRISTI NAVAL AIR STATION: | | | |
| GENERAL PURPOSE WAREHOUSE | 79,300 | 79,300 | ............ |
| SAN ANTONIO: | | | |
| NSA/CSS TEXAS CRYPTOLOGIC CENTER (INC) | 152,000 | 152,000 | ............ |
| AIR NATIONAL GUARD: | | | |
| FORT WORTH: | | | |
| C–130J ADAL FUEL CELL BUILDING 1674 | 13,100 | 13,100 | ............ |
| TOTAL, TEXAS | 486,700 | 486,700 | ............ |
| UTAH | | | |
| AIR FORCE: | | | |
| HILL AIR FORCE BASE: | | | |
| T–7A DEPOT MAINTENANCE COMPLEX, INC | 50,000 | 50,000 | ............ |
| ARMY NATIONAL GUARD: | | | |
| NEPHI READINESS CENTER: | | | |
| NATIONAL GUARD VEHICLE MAINTENANCE SHOP | 20,000 | 20,000 | ............ |
| TOTAL, UTAH | 70,000 | 70,000 | ............ |

## Military Construction Project Listing by Location (Continued)

| Installation and project | Budget estimate | Committee recommendation | Committee recommendation compared to budget estimate |
|---|---|---|---|
| VERMONT | | | |
| ARMY NATIONAL GUARD: | | | |
| ST ALBANS: | | | |
| READINESS CENTER: DESIGN | | 4,030 | +4,030 |
| TOTAL, VERMONT | | 4,030 | +4,030 |
| VIRGINIA | | | |
| ARMY: | | | |
| JOINT BASE MYER–HENDERSON: | | | |
| BARRACKS | 180,000 | 180,000 | |
| HORSE FARM LAND ACQUISITION | 8,500 | | −8,500 |
| NAVY AND MARINE CORPS: | | | |
| MARINE CORPS BASE QUANTICO: | | | |
| CHILD DEVELOPMENT CENTER: DESIGN | | 8,200 | +8,200 |
| NAVAL AIR STATION OCEANA: | | | |
| UNACCOMPANIED HOUSING: DESIGN | | 16,000 | +16,000 |
| NORFOLK NAVAL SHIPYARD: | | | |
| CHILD DEVELOPMENT CENTER: DESIGN | | 1,400 | +1,400 |
| PORTSMOUTH: | | | |
| DRY DOCK 3 MODERNIZATION (INC) | 54,366 | 54,366 | |
| YORKTOWN: | | | |
| CONTAINERIZED LONG WEAPONS STORAGE MAGAZINE | 52,610 | 52,610 | |
| CONVENTIONAL PROMPT STRIKE TEST FACILITY | 47,130 | 47,130 | |
| CPS WEAPONS MAINTENANCE, OPS & STORAGE FAC | 52,110 | 52,110 | |
| AIR FORCE: | | | |
| JOINT BASE LANGLEY–EUSTIS: | | | |
| DORMITORY | 81,000 | 81,000 | |
| DEFENSE-WIDE: | | | |
| FORT BELVOIR: | | | |
| DEFENSE HEALTH HEADQUARTERS | 225,000 | 225,000 | |
| JOINT EXPEDITIONARY BASE LITTLE CREEK—FT STORY: | | | |
| SOF HUMAN PERFORMANCE TRAINING CENTER | 32,000 | 32,000 | |
| PENTAGON: | | | |
| METRO ENTRANCE PEDESTRIAN ACCESS CONTROL PT | 36,800 | 36,800 | |
| ARMY RESERVE: | | | |
| RICHMOND: | | | |
| AREA MAINTENANCE SUPPORT ACTIVITY / VMS | 23,000 | 23,000 | |
| TOTAL, VIRGINIA | 792,516 | 809,616 | +17,100 |
| WASHINGTON | | | |
| ARMY: | | | |
| JOINT BASE LEWIS–MCCHORD: | | | |
| BARRACKS | 161,000 | 161,000 | |
| SUPPLY SUPPORT ACTIVITY | 31,000 | 31,000 | |
| YAKIMA TRAINING CENTER: | | | |
| WASTEWATER TREATMENT PLANT: DESIGN | | 6,460 | +6,460 |
| NAVY AND MARINE CORPS: | | | |
| BANGOR: | | | |
| LAUNCHER EQUIPMENT PROCESSING BUILDING | 200,550 | 200,550 | |
| PUGET SOUND: | | | |
| CVN 78 AIRCRAFT CARRIER ELECT UPGRADES | 182,200 | 182,200 | |
| AIR FORCE: | | | |
| FAIRCHILD AIR FORCE BASE: | | | |
| KC–135 ADAL AIRCRAFT PARTS WAREHOUSE: DESIGN | | 2,500 | +2,500 |
| TAXILANE PULL–THROUGH SPOTS: DESIGN | | 4,500 | +4,500 |

| Installation and project | Budget estimate | Committee recommendation | Committee recommendation compared to budget estimate |
|---|---|---|---|
| DEFENSE-WIDE: | | | |
| KEYPORT: | | | |
| SOF COLDWATER TRAINING/AUSTERE ENVIRON FAC | 35,000 | | −35,000 |
| WHIDBEY ISLAND: | | | |
| HYDRANT FUELING SYSTEM | 54,000 | 54,000 | |
| ARMY NATIONAL GUARD: | | | |
| CAMP MURRAY: | | | |
| NATIONAL GUARD/RESERVE CENTER BUILDING | 40,000 | 40,000 | |
| TACOMA: | | | |
| CONTROLLED HUMIDITY WAREHOUSE: MINOR CONSTRUCTION | | 6,000 | +6,000 |
| AIR NATIONAL GUARD: | | | |
| CAMP MURRAY: | | | |
| NATIONAL GUARD/RESERVE CENTER (ANG COST SHARE): MINOR CONSTRUCTION | | 5,700 | +5,700 |
| NAVY RESERVE: | | | |
| JOINT BASE LEWIS–MCCHORD: | | | |
| PARACHUTE SURVIVAL TRAINING FACILITY | 26,610 | 26,610 | |
| TOTAL, WASHINGTON | 730,360 | 720,520 | −9,840 |
| WEST VIRGINIA | | | |
| ARMY NATIONAL GUARD: | | | |
| BLUEFIELD: | | | |
| NATIONAL GUARD READINESS CENTER: DESIGN | | 4,000 | +4,000 |
| CAMP DAWSON: | | | |
| FIRE STATION SUPPORT BUILDING: MINOR CONSTRUCTION | | 6,000 | +6,000 |
| CAMP DAWSON—KINGWOOD: | | | |
| GENERAL INSTRUCTION BUILDING: DESIGN | | 1,200 | +1,200 |
| JOINT FORCE HEADQUARTERS—CHARLESTON: | | | |
| NATIONAL GUARD/RESERVE CENTER BUILDING: DESIGN | | 1,400 | +1,400 |
| PHYSICAL FITNESS CENTER (MULTIPURPOSE BUILDING): DESIGN | | 2,500 | +2,500 |
| TOTAL, WEST VIRGINIA | | 15,100 | +15,100 |
| WISCONSIN | | | |
| ARMY NATIONAL GUARD: | | | |
| BLACK RIVER FALLS: | | | |
| NATIONAL GUARD READINESS CENTER: DESIGN | | 2,600 | +2,600 |
| WAUSAU: | | | |
| NATIONAL GUARD READINESS CENTER: DESIGN | | 7,150 | +7,150 |
| WISCONSIN RAPIDS: | | | |
| NATIONAL GUARD READINESS CENTER: DESIGN | | 4,940 | +4,940 |
| ARMY RESERVE: | | | |
| MANITOWOC: | | | |
| ARMY RESERVE VEHICLE MAINTENANCE SHOP: DESIGN | | 150 | +150 |
| TOTAL, WISCONSIN | | 14,840 | +14,840 |
| WYOMING | | | |
| AIR FORCE: | | | |
| FE WARREN AIR FORCE BASE: | | | |
| GBSD CONSOLIDATED MAINTENANCE FACILITY | 194,000 | 194,000 | |
| GBSD LAND ACQUISITION PHASE 2 | 139,000 | 139,000 | |
| GBSD UTILITY CORRIDOR, INC | 70,000 | 70,000 | |

## Military Construction Project Listing by Location (Continued)

| Installation and project | Budget estimate | Committee recommendation | Committee recommendation compared to budget estimate |
|---|---|---|---|
| TOTAL, WYOMING | 403,000 | 403,000 | |
| AUSTRALIA | | | |
| NAVY AND MARINE CORPS: | | | |
| DARWIN: | | | |
| PDI: AIRCRAFT MAINTENANCE HANGAR | 117,380 | 117,380 | |
| PDI: MAINTENANCE SUPPORT FACS | 62,320 | 62,320 | |
| TOTAL, AUSTRALIA | 179,700 | 179,700 | |
| BELGIUM | | | |
| ARMY: | | | |
| SHAPE HQ: | | | |
| YOUTH CENTER | 45,000 | 45,000 | |
| TOTAL, BELGIUM | 45,000 | 45,000 | |
| CUBA | | | |
| DEFENSE-WIDE: | | | |
| GUANTANAMO BAY NAVAL STATION: | | | |
| AMBULATORY CARE CENTER REPLACEMENT INCR 2 | 96,829 | 96,829 | |
| TOTAL, CUBA | 96,829 | 96,829 | |
| DENMARK | | | |
| AIR FORCE: | | | |
| KARUP: | | | |
| EDI: DABS–FEV STORAGE | 110,000 | 110,000 | |
| TOTAL, DENMARK | 110,000 | 110,000 | |
| FEDERATED STATES OF MICRONESIA | | | |
| AIR FORCE: | | | |
| YAP AIRFIELD: | | | |
| PDI: RUNWAY EXTENSION, INC | 96,000 | 96,000 | |
| TOTAL, FEDERATED STATES OF MICRONESIA | 96,000 | 96,000 | |
| GERMANY | | | |
| ARMY: | | | |
| ANSBACH: | | | |
| BARRACKS | 100,000 | 100,000 | |
| BARRACKS | 91,000 | 91,000 | |
| HOHENFELS: | | | |
| SIMULATIONS CENTER: COST TO COMPLETE | 35,000 | 35,000 | |
| SMITH BARRACKS: | | | |
| BARRACKS | 61,000 | 61,000 | |
| WIESBADEN MIL CMTY: | | | |
| CHILD DEVELOPMENT CENTER | 44,000 | 44,000 | |
| DEFENSE-WIDE: | | | |
| SPANGDAHLEM AIR BASE: | | | |
| SPANGDAHLEM ELEM SCHOOL REPLACE: COST TO COMPLETE | 6,500 | 6,500 | |

| Installation and project | Budget estimate | Committee recommendation | Committee recommendation compared to budget estimate |
|---|---|---|---|
| TOTAL, GERMANY | 337,500 | 337,500 | .................. |
| GUAM | | | |
| NAVY AND MARINE CORPS: | | | |
| JOINT REGION MARIANAS: | | | |
| PDI: YOUTH CENTER | 78,730 | 78,730 | .................. |
| PDI: EARTH COVERED MAGAZINES | 107,439 | 107,439 | .................. |
| DEFENSE-WIDE: | | | |
| JOINT REGION MARIANAS: | | | |
| GUAM HIGH SCHOOL TEMPORARY FACILITIES | 26,000 | 26,000 | .................. |
| PDI: GDS, COMMAND CENTER (INC) | 187,212 | 147,212 | −40,000 |
| PDI: GDS, EIAMD, PH1 (INC) | 278,267 | 238,267 | −40,000 |
| TOTAL, GUAM | 677,648 | 597,648 | −80,000 |
| JAPAN | | | |
| AIR FORCE: | | | |
| KADENA AIR BASE: | | | |
| PDI: THEATER A/C CORROSION CONTROL CTR, INC 3 | 132,700 | 132,700 | .................. |
| DEFENSE-WIDE: | | | |
| CAMP BUTLER: | | | |
| KUBASAKI HIGH SCHOOL | 160,000 | 160,000 | .................. |
| YOKOSUKA: | | | |
| KINNICK HIGH SCHOOL INC | 40,386 | 40,386 | .................. |
| TOTAL, JAPAN | 333,086 | 333,086 | .................. |
| KOREA | | | |
| DEFENSE-WIDE: | | | |
| KUNSAN AIR BASE: | | | |
| AMBULATORY CARE CENTER REPLACEMENT | 64,942 | 64,942 | .................. |
| TOTAL, KOREA | 64,942 | 64,942 | .................. |
| PUERTO RICO | | | |
| ARMY RESERVE: | | | |
| FORT BUCHANAN: | | | |
| ADVANCED SKILLS TRAINING BARRACKS | 39,000 | 39,000 | .................. |
| TOTAL, PUERTO RICO | 39,000 | 39,000 | .................. |
| SPAIN | | | |
| AIR FORCE: | | | |
| ROTA: | | | |
| NATO STRATEGIC AIRLIFT HANGAR | 15,200 | 15,200 | .................. |
| TOTAL, SPAIN | 15,200 | 15,200 | .................. |
| UNITED KINGDOM | | | |
| AIR FORCE: | | | |
| ROYAL AIR FORCE LAKENHEATH: | | | |
| SURETY: BARRIER SYSTEMS | 185,000 | 185,000 | .................. |
| ROYAL AIR FORCE MILDENHALL: | | | |
| SOW CAMPUS INFRASTRUCTURE | 51,000 | 51,000 | .................. |

## Military Construction Project Listing by Location (Continued)

| Installation and project | Budget estimate | Committee recommendation | Committee recommendation compared to budget estimate |
|---|---|---|---|
| DEFENSE-WIDE: | | | |
| ROYAL AIR FORCE LAKENHEATH: | | | |
| LAKENHEATH HIGH SCHOOL | 153,000 | 153,000 | |
| TOTAL, UNITED KINGDOM | 389,000 | 389,000 | |
| WORLDWIDE UNSPECIFIED | | | |
| ARMY: | | | |
| DESIGN | 273,727 | 297,727 | +24,000 |
| HOST NATION SUPPORT | 25,000 | 25,000 | |
| EDI: MINOR CONSTRUCTION | 14,519 | 14,519 | |
| MINOR CONSTRUCTION | 97,000 | 117,000 | +20,000 |
| PDI: DESIGN | 26,011 | 26,011 | |
| PDI: INDOPACOM MINOR CONSTRUCTION PILOT | 66,600 | 66,600 | |
| PDI: MINOR CONSTRUCTION | 8,000 | 8,000 | |
| SUBTOTAL, ARMY | 510,857 | 554,857 | +44,000 |
| NAVY AND MARINE CORPS: | | | |
| DESIGN | 797,446 | 797,446 | |
| DPRI UNSPECIFIED MINOR CONSTRUCTION | 21,302 | 21,302 | |
| MINOR CONSTRUCTION | 77,794 | 77,794 | |
| MINOR CONSTRUCTION | 124,524 | 124,524 | |
| SUBTOTAL, NAVY AND MARINE CORPS | 1,021,066 | 1,021,066 | |
| AIR FORCE: | | | |
| DESIGN | 355,011 | 358,011 | +3,000 |
| DESIGN | 84,915 | 84,915 | |
| MINOR CONSTRUCTION | 104,700 | 104,700 | |
| MINOR CONSTRUCTION | 24,900 | 24,900 | |
| SUBTOTAL, AIR FORCE | 569,526 | 572,526 | +3,000 |
| DEFENSE-WIDE: | | | |
| ENERGY RESILIENCE AND CONSERV INVEST PROG | 636,000 | 636,000 | |
| DESIGN: | | | |
| INDOPACOM | | 30,000 | +30,000 |
| ERCIP DESIGN | 96,238 | 96,238 | |
| MISSILE DEFENSE AGENCY | 4,745 | 84,745 | +80,000 |
| DEFENSE HEALTH AGENCY | 46,751 | 46,751 | |
| DEFENSE LOGISTICS AGENCY | 105,000 | 105,000 | |
| DEPARTMENT OF DEFENSE EDUCATION ACTIVITY | 7,501 | 7,501 | |
| WASHINGTON HEADQUARTERS SERVICE | 1,508 | 1,508 | |
| THE JOINT STAFF | 1,964 | 1,964 | |
| US SPECIAL OPERATIONS COMMAND | 35,495 | 35,495 | |
| NATIONAL SECURITY AGENCY | 41,928 | 41,928 | |
| DEFENSE-WIDE | 26,081 | 26,081 | |
| MINOR CONSTRUCTION: | | | |
| THE JOINT STAFF | 11,146 | 11,146 | |
| MISSILE DEFENSE AGENCY | 5,277 | 5,277 | |
| DEFENSE LOGISTICS AGENCY | 13,333 | 13,333 | |
| DEFENSE HEALTH AGENCY | 18,000 | 18,000 | |
| DEPARTMENT OF DEFENSE EDUCATION ACTIVITY | 7,400 | 7,400 | |
| DEFENSE-WIDE | 3,000 | 3,000 | |
| US SPECIAL OPERATIONS COMMAND | 24,109 | 24,109 | |
| NATIONAL SECURITY AGENCY | 6,000 | 6,000 | |

| Installation and project | Budget estimate | Committee recommendation | Committee recommendation compared to budget estimate |
|---|---|---|---|
| SUBTOTAL, DEFENSE-WIDE | 1,091,476 | 1,201,476 | +110,000 |
| ARMY NATIONAL GUARD: | | | |
| DESIGN | 25,529 | 40,529 | +15,000 |
| MINOR CONSTRUCTION | 45,000 | 45,000 | |
| SUBTOTAL, ARMY NATIONAL GUARD | 70,529 | 85,529 | +15,000 |
| AIR NATIONAL GUARD: | | | |
| DESIGN | 10,792 | 14,792 | +4,000 |
| MINOR CONSTRUCTION | 40,200 | 40,200 | |
| SUBTOTAL, AIR NATIONAL GUARD | 50,992 | 54,992 | +4,000 |
| ARMY RESERVE: | | | |
| DESIGN | 31,508 | 31,508 | |
| MINOR CONSTRUCTION | 3,524 | 3,524 | |
| SUBTOTAL, ARMY RESERVE | 35,032 | 35,032 | |
| NAVY RESERVE: | | | |
| DESIGN | 663 | 663 | |
| DESIGN | 2,556 | 2,556 | |
| SUBTOTAL, NAVY RESERVE | 3,219 | 3,219 | |
| AIR FORCE RESERVE: | | | |
| DESIGN | 562 | 11,562 | +11,000 |
| MINOR CONSTRUCTION | 701 | 11,701 | +11,000 |
| SUBTOTAL, AIR FORCE RESERVE | 435,127 | 457,127 | +22,000 |
| NORTH ATLANTIC TREATY ORGANIZATION SECURITY INVESTMENT PROGRAM | | | |
| NATO SECURITY INVESTMENT PROGRAM | 433,864 | 433,864 | |
| BASE REALIGNMENT AND CLOSURE | | | |
| DEPARTMENT OF DEFENSE BASE CLOSURE ACCOUNT | | | |
| ARMY | 212,556 | 237,556 | +25,000 |
| NAVY | 111,697 | 111,697 | |
| AIR FORCE | 121,952 | 146,952 | +25,000 |
| DEFENSE-WIDE | 1,756 | 1,756 | |
| TOTAL, BASE REALIGNMENT AND CLOSURE | 447,961 | 497,961 | +50,000 |
| SUBTOTAL, WORLDWIDE UNSPECIFIED | 4,235,785 | 4,483,785 | +248,000 |
| FAMILY HOUSING, ARMY | | | |
| GEORGIA: | | | |
| FORT EISENHOWER: | | | |
| FORT EISENHOWER MHPI EQUITY INVESTMENT | 50,000 | 50,000 | |
| BELGIUM: | | | |
| CHIEVRES AIR BASE: | | | |
| FAMILY HOUSING NEW CONSTRUCTION (84 UNITS) | 100,954 | 100,954 | |
| GERMANY: | | | |
| BAUMHOLDER: | | | |
| FAMILY HOUSING REPLACEMENT CONSTR (54 UNITS) | 63,246 | 63,246 | |

## Military Construction Project Listing by Location (Continued)

| Installation and project | Budget estimate | Committee recommendation | Committee recommendation compared to budget estimate |
|---|---|---|---|
| JAPAN: | | | |
| SAGAMIHARA: | | | |
| FAMILY HOUSING IMPROVEMENTS CONSTR (35 UNITS) | 31,114 | 31,114 | |
| DESIGN | 31,333 | 31,333 | |
| SUBTOTAL, CONSTRUCTION | 276,647 | 276,647 | |
| OPERATION AND MAINTENANCE: | | | |
| FURNISHINGS | 18,065 | 18,065 | |
| LEASED HOUSING | 129,703 | 129,703 | |
| MAINTENANCE OF REAL PROPERTY FACILITIES | 127,097 | 127,097 | |
| MANAGEMENT ACCOUNT | 62,060 | 62,060 | |
| MILITARY HOUSING PRIVITIZATION INITIATIVE | 69,579 | 69,579 | |
| MISCELLANEOUS | 357 | 357 | |
| SERVICES | 8,273 | 8,273 | |
| UTILITIES | 60,477 | 60,477 | |
| SUBTOTAL, OPERATIONS AND MAINTENANCE | 475,611 | 475,611 | |
| FAMILY HOUSING, NAVY AND MARINE CORPS | | | |
| GUAM: | | | |
| NAVAL SUPPORT ACTIVITY ANDERSEN: | | | |
| REPLACE ANDERSEN HOUSING PH 9 (136 UNITS) | 103,863 | 103,863 | |
| REPLACE ANDERSEN HOUSING PH 10 (42 UNITS) | 93,112 | 93,112 | |
| JAPAN: | | | |
| MARINE CORPS AIR STATION IWAKUNI: | | | |
| CONSTRUCTION IMPROVEMENTS | 35,438 | 35,438 | |
| DESIGN | 13,329 | 13,329 | |
| SUBTOTAL, CONSTRUCTION | 245,742 | 245,742 | |
| OPERATION AND MAINTENANCE: | | | |
| FURNISHINGS | 16,839 | 16,839 | |
| HOUSING PRIVATIZATION SUPPORT | 60,283 | 60,283 | |
| LEASING | 67,412 | 67,412 | |
| MAINTENANCE | 109,504 | 109,504 | |
| MANAGEMENT | 61,240 | 61,240 | |
| MISCELLANEOUS | 427 | 427 | |
| SERVICES | 17,332 | 17,332 | |
| UTILITIES | 44,180 | 44,180 | |
| SUBTOTAL, OPERATION AND MAINTENANCE | 377,217 | 377,217 | |
| FAMILY HOUSING, AIR FORCE | | | |
| ALASKA: | | | |
| JOINT BASE ELMENDORF-RICHARDSON: | | | |
| MHPI RESTRUCTURE—JBER PHASE III | 120,000 | 120,000 | |
| TEXAS: | | | |
| LACKLAND AIR FORCE BASE: | | | |
| MHPI RESTRUCTURE—LACKLAND | 24,000 | 24,000 | |
| GERMANY: | | | |
| RAMSTEIN AIR BASE: | | | |
| CONSTRUCT 2 GOQ UNITS | 4,350 | 4,350 | |
| KMC 02—CONSTRUCT TWO CAR GARAGES (5 UNITS) | 1,400 | 1,400 | |
| JAPAN: | | | |
| YOKOTA AIR BASE: | | | |
| CONST—IMP—YOKOTA PAIP 8B WEST (19 UNITS) | 26,242 | 26,242 | |
| CONST—IMP—YOKOTA PAIP 9 PHASE 2 (32 UNITS) | 39,000 | 39,000 | |

| Installation and project | Budget estimate | Committee recommendation | Committee recommendation compared to budget estimate |
|---|---|---|---|
| DESIGN | 6,557 | 6,557 | |
| SUBTOTAL, CONSTRUCTION | 221,549 | 221,549 | |
| OPERATION AND MAINTENANCE: | | | |
| FURNISHINGS | 24,230 | 24,230 | |
| HOUSING PRIVATIZATION | 32,508 | 32,508 | |
| LEASING | 6,278 | 6,278 | |
| MAINTENANCE | 127,023 | 127,023 | |
| MANAGEMENT | 71,384 | 71,384 | |
| MISCELLANEOUS | 2,426 | 2,426 | |
| SERVICES | 12,446 | 12,446 | |
| UTILITIES | 49,955 | 49,955 | |
| SUBTOTAL, OPERATION AND MAINTENANCE | 326,250 | 326,250 | |
| FAMILY HOUSING, DEFENSE-WIDE | | | |
| OPERATION AND MAINTENANCE: | | | |
| DEFENSE INTELLIGENCE AGENCY: | | | |
| FURNISHINGS | 687 | 687 | |
| LEASING | 32,983 | 32,983 | |
| UTILITIES | 4,358 | 4,358 | |
| NATIONAL SECURITY AGENCY: | | | |
| FURNISHINGS | 91 | 91 | |
| LEASING | 13,986 | 13,986 | |
| MAINTENANCE | 36 | 36 | |
| UTILITIES | 15 | 15 | |
| SUBTOTAL, OPERATION AND MAINTENANCE | 52,156 | 52,156 | |
| DOD FAMILY HOUSING IMPROVEMENT FUND | 8,195 | 8,195 | |
| DOD MILITARY UNACCOMPANIED HOUSING IMPROVEMENT FUND | 497 | 497 | |
| ADMINISTRATIVE PROVISIONS | | | |
| UNFUNDED REQUIREMENTS (SEC 124): | | | |
| MILITARY CONSTRUCTION, ARMY | | 26,500 | +26,500 |
| MILITARY CONSTRUCTION, NAVY AND MARINE CORPS | | 50,400 | +50,400 |
| MILITARY CONSTRUCTION, AIR FORCE | | 183,000 | +183,000 |
| MILITARY CONSTRUCTION, DEFENSE-WIDE | | 134,270 | +134,270 |
| MILITARY CONSTRUCTION, ARMY NATIONAL GUARD | | 54,251 | +54,251 |
| MILITARY CONSTRUCTION, AIR NATIONAL GUARD | | 6,000 | +6,000 |
| MILITARY CONSTRUCTION, ARMY RESERVE | | 63,000 | +63,000 |
| MILITARY INSTALLATION RESILIENCE (SEC 128): | | | |
| MILITARY CONSTRUCTION, ARMY | | 10,000 | +10,000 |
| MILITARY CONSTRUCTION, NAVY AND MARINE CORPS | | 10,000 | +10,000 |
| MILITARY CONSTRUCTION, AIR FORCE | | 10,000 | +10,000 |
| FOREIGN MILITARY TRAINING INFRASTRUCTURE (SEC 129): | | | |
| MILITARY CONSTRUCTION, AIR FORCE | | 100,000 | +100,000 |
| MILITARY CONSTRUCTION, DEFENSE-WIDE | | 2,500 | +2,500 |
| MILITARY CONSTRUCTION, AIR NATIONAL GUARD | | 26,000 | +26,000 |
| AUTHORIZED PROJECTS (SEC 131): | | | |
| MILITARY CONSTRUCTION, AIR NATIONAL GUARD | | 32,000 | +32,000 |
| TOTAL, ADMINISTRATIVE PROVISIONS | | 707,921 | +707,921 |
| RECAP | | | |
| ARMY | 2,311,157 | 2,380,477 | +69,320 |
| NAVY AND MARINE CORPS | 4,540,899 | 4,874,699 | +333,800 |

## Military Construction Project Listing by Location (Continued)

| Installation and project | Budget estimate | Committee recommendation | Committee recommendation compared to budget estimate |
|---|---|---|---|
| AIR FORCE | 3,187,126 | 3,549,626 | +362,500 |
| DEFENSE-WIDE | 3,733,163 | 3,601,163 | −132,000 |
| ARMY NATIONAL GUARD | 362,129 | 456,459 | +94,330 |
| AIR NATIONAL GUARD | 190,792 | 307,792 | +117,000 |
| ARMY RESERVE | 255,032 | 375,682 | +120,650 |
| NAVY RESERVE | 29,829 | 29,829 | |
| AIR FORCE RESERVE | 69,263 | 107,663 | +38,400 |
| NATO | 433,864 | 433,864 | |
| DEPARTMENT OF DEFENSE BASE CLOSURE ACCOUNT (BRAC) | 447,961 | 497,961 | +50,000 |
| FAMILY HOUSING, ARMY | 752,258 | 752,258 | |
| CONSTRUCTION | (276,647) | (276,647) | |
| OPERATION AND MAINTENANCE | (475,611) | (475,611) | |
| FAMILY HOUSING, NAVY AND MARINE CORPS | 622,959 | 622,959 | |
| CONSTRUCTION | (245,742) | (245,742) | |
| OPERATION AND MAINTENANCE | (377,217) | (377,217) | |
| FAMILY HOUSING, AIR FORCE | 547,799 | 547,799 | |
| CONSTRUCTION | (221,549) | (221,549) | |
| OPERATION AND MAINTENANCE | (326,250) | (326,250) | |
| FAMILY HOUSING, DEFENSE-WIDE | 52,156 | 52,156 | |
| OPERATION AND MAINTENANCE | (52,156) | (52,156) | |
| DOD FAMILY HOUSING IMPROVEMENT FUND | 8,195 | 8,195 | |
| DOD UNACCOMPANIED HOUSING IMPROVEMENT FUND | 497 | 497 | |
| ADMINISTRATIVE PROVISIONS | | 707,921 | +707,921 |
| GRAND TOTAL | 17,545,079 | 19,307,000 | +1,761,921 |

## Disclosure of Congressionally Directed Spending Items

The Constitution vests in the Congress the power of the purse. The Committee believes strongly that Congress should make the decisions on how to allocate the people's money.

As defined in Rule XLIV of the Standing Rules of the Senate, the term ''congressionally directed spending item'' means a provision or report language included primarily at the request of a Senator, providing, authorizing, or recommending a specific amount of discretionary budget authority, credit authority, or other spending authority for a contract, loan, loan guarantee, grant, loan authority, or other expenditure with or to an entity, or targeted to a specific State, locality or congressional district, other than through a statutory or administrative, formula-driven, or competitive award process.

For each item, a Member is required to provide a certification that neither the Member nor the Member's immediate family has a pecuniary interest in such congressionally directed spending item. Such certifications are available to the public on the website of the Senate Committee on Appropriations (https:// www.appropriations.senate.gov/congressionally-directed-spending-requests).

Following is a list of congressionally directed spending items included in the Senate recommendation discussed in this report, along with the name of each Senator who submitted a request to the Committee of jurisdiction for each item so identified. Neither the Committee recommendation nor this report contains any limited tax benefits or limited tariff benefits as defined in rule XLIV.

## Congressionally Directed Spending Items

[Dollars in thousands]

| Account | State | Location | Project | Amount | Requestor(s) |
|---|---|---|---|---|---|
| Army | Alabama | Fort Novosel | Crash/Rescue Station at Hatch Army Heliport: Minor Construction | $6,500 | Britt/Tuberville |
| Army | Alabama | Fort Novosel | Crash/Rescue Station at Tabernacle Stagefield: Minor Construction. | 6,600 | Britt |
| Army | Alabama | Fort Novosel | Flat Iron Support Facility: Minor Construction | 5,500 | Britt/Tuberville |
| Army | Alabama | Fort Novosel | Flight Control Tower—Hanchey Army Airfield: Design | 1,300 | Britt |
| Army | Alabama | Fort Novosel | Installation Service Node Facility: Design | 2,600 | Britt |
| Air Force | Alaska | Eielson Air Force Base | Joint Mobility Center Expansion: Design | 5,800 | Murkowski |
| Air Force | Alaska | Eielson Air Force Base | Permanent Party Dormitory: Design | 6,700 | Murkowski |
| Air Force | Alaska | Joint Base Elmendorf-Richardson | Precision Guided Munitions Complex: Design | 15,600 | Murkowski |
| Army | Alaska | Fort Wainwright | Dining Facility: Design | 7,700 | Murkowski |
| Army | Alaska | Fort Wainwright | Tactical Equipment Maintenance Facility: Design | 10,800 | Murkowski |
| Air National Guard | Arizona | Morris Air National Guard Base | Base Entry Complex: Cost to Complete | 7,000 | Kelly/Sinema |
| Army | Arizona | Yuma Proving Ground | Radar Operations & Maintenance Facility: Minor Construction | 6,000 | Kelly/Sinema |
| Navy and Marine Corps | Arizona | Marine Corps Air Station Yuma | UDP Transient Barracks: Design | 3,800 | Kelly/Sinema |
| Navy and Marine Corps | Arizona | Marine Corps Air Station Yuma | Water Treatment Plant | 90,000 | Kelly/Sinema |
| Air Force | California | Beale Air Force Base | Multi-Domain Operations Complex | 44,000 | Butler/Padilla |
| Army Reserve | California | Bell | Army Reserve Center: Design | 3,750 | Butler |
| Air Force | Colorado | Buckley Space Force Base | Power Independence | 60,000 | Bennet/Hickenlooper |
| Army National Guard | Connecticut | Groton | Aircraft Maintenance Hangar: Design | 6,500 | Blumenthal/Murphy |
| Navy and Marine Corps | Connecticut | Naval Submarine Base New London | Submarine Storage, Maintenance & Operations Facility: Design | 3,200 | Blumenthal/Murphy |
| Air National Guard | Delaware | New Castle County Airport | Entry Control Facility: Minor Construction | 8,000 | Carper/Coons |
| Army National Guard | Delaware | Dagsboro | Dagsboro Readiness Center: Minor Construction | 7,500 | Carper/Coons |
| Army | Georgia | Fort Stewart | Enlisted Unaccompanied Personnel Barracks (PN 96725): Design | 10,500 | Ossoff/Warnock |
| Army | Georgia | Fort Stewart/Hunter Army Airfield | SOF Military Working Dog Kennel Facility: Design | 1,230 | Ossoff |
| Army Reserve | Georgia | Marine Corps Logistics Base Albany | Army Reserve Center | 40,000 | Ossoff/Warnock |
| Air Force | Hawaii | Joint Base Pearl Harbor-Hickam | General Purpose Maintenance Hangar: Design | 7,000 | Hirono/Schatz |
| Air Force | Hawaii | Joint Base Pearl Harbor-Hickam | Mobility Processing Center/War Reserve Material Warehouse: Design. | 11,000 | Hirono/Schatz |
| Army | Hawaii | Pohakuloa Training Area | Airfield Operations Building | 20,000 | Hirono/Schatz |
| Navy and Marine Corps | Hawaii | Joint Base Pearl Harbor-Hickam | (Red Hill) Water Treatment Plant | 90,000 | Hirono/Schatz |
| Army | Illinois | Rock Island Arsenal | Forging Equipment Annex: Minor Construction | 5,000 | Durbin |
| Army | Illinois | Rock Island Arsenal | New Weapons Quality Assurance/Calibration Facility: Minor Construction. | 4,250 | Durbin |
| Army National Guard | Illinois | Marseilles Training Center | Marseilles Training Center Range Renovation: Minor Construction | 4,300 | Duckworth/Durbin |
| Army | Kansas | Fort Leavenworth | B95 Fire Station Truck Bay Addition: Minor Construction | 8,200 | Moran |

| | | | | | |
|---|---|---|---|---|---|
| Army | Kansas | Fort Leavenworth | Fire Station Headquarters: Design | 1,800 | Moran |
| Army | Kansas | Fort Leavenworth | Functional Fitness Center: Design | 1,700 | Moran |
| Air National Guard | Kentucky | Louisville International Airport-Standiford Fld. | C–130J ADAL Fuel Cell Hangar: Minor Construction | 8,500 | McConnell |
| Army | Kentucky | Blue Grass Army Depot | Military Container Production Facility: Design | 27,000 | McConnell |
| Army | Kentucky | Fort Campbell | Foundry Training Facility: Minor Construction | 8,100 | McConnell |
| Army | Kentucky | Fort Campbell | Modernized Hangar: Design | 11,000 | McConnell |
| Army Reserve | Kentucky | Fort Knox | Aviation Support Facility | 75,000 | McConnell |
| Air Force | Louisiana | Barksdale Air Force Base | Child Development Center | 22,000 | Cassidy/Kennedy |
| Army | Louisiana | Fort Johnson | Rotational Unit Billeting Area: Design | 9,900 | Cassidy |
| Air National Guard | Maine | Bangor International Airport | Aircraft Maintenance Squadron/Aircraft Ground Equipment Facility: Design. | 800 | Collins/King |
| Air National Guard | Maine | Bangor International Airport | Fuel Cell Hangar | 50,000 | Collins |
| Army National Guard | Maine | Saco | Southern Maine Readiness Center: Design | 800 | Collins/King |
| Army National Guard | Maine | Woodville | Automated Qualification Training Range: Design | 1,000 | Collins/King |
| Navy and Marine Corps | Maine | Portsmouth Naval Shipyard | Parking Structure: Design | 9,500 | Collins/King/Shaheen |
| Army National Guard | Maryland | Hagerstown | Field Maintenance Shop: Design | 2,470 | Cardin/Van Hollen |
| Navy and Marine Corps | Maryland | Naval Surface Warfare Center Indian Head | Contained Burn Facility | 50,000 | Cardin/Van Hollen |
| Army National Guard | Michigan | Olympia Armory | Readiness Center Addition/Alteration: Design | 4,400 | Peters/Stabenow |
| Air National Guard | Minnesota | Minneapolis St Paul International Airport | C–130J Fuel Cell Maintenance Hangar: Design | 4,300 | Klobuchar/Smith |
| Air Force | Mississippi | Keesler Air Force Base | Air Traffic Control Tower | 25,000 | Hyde-Smith/Wicker |
| Air National Guard | Mississippi | Key Field | Fuel Cell/Corrosion Control Hangar: Design | 5,000 | Hyde-Smith/Wicker |
| Air National Guard | Mississippi | Key Field | Maintenance Hangar: Design | 5,000 | Hyde-Smith |
| Army | Mississippi | Army Engineer Research and Development Center. | Jet Engine Thermal Simulator: Design | 700 | Hyde-Smith/Wicker |
| Defense-wide (DLA) | Mississippi | Key Field | Upgrade Fuel Hydrant System: Design | 1,000 | Hyde-Smith |
| Air Force | Nebraska | Offutt Air Force Base | Consolidated Training Complex/Professional Development Center: Design. | 6,000 | Fischer |
| Air Force | Nebraska | Offutt Air Force Base | Survivable Airborne Operations Center Complex: Design | 70,000 | Fischer |
| Army National Guard | Nebraska | Hastings | Collective Training Unaccompanied Housing: Minor Construction | 5,900 | Fischer |
| Air Force | Nevada | Nellis Air Force Base | Regional Communications Center: Design | 3,500 | Cortez Masto/Rosen |
| Air National Guard | Nevada | Reno-Tahoe International Airport | Aircraft Parking Ramp Extension: Design | 1,000 | Cortez Masto/Rosen |
| Air National Guard | Nevada | Reno-Tahoe International Airport | Aircraft Support Equipment and Engine Facility: Design | 2,000 | Cortez Masto/Rosen |
| Air National Guard | Nevada | Reno-Tahoe International Airport | Fuel Cell Hangar: Design | 3,000 | Cortez Masto/Rosen |
| Army National Guard | New Hampshire | Concord | Lakota Hangar Addition and Environmental Upgrades: Design | 1,200 | Shaheen |
| Army National Guard | New Hampshire | Pembrooke | Readiness Center Addition/Alteration: Minor Construction | 6,500 | Shaheen |
| Air National Guard | New Jersey | Atlantic City International Airport | Fighter Operations Facility: Design | 3,400 | Booker |
| Air Force | New Mexico | Cannon Air Force Base | ADAL Security Forces Facility: Minor Construction | 4,000 | Heinrich |
| Air Force | New Mexico | Cannon Air Force Base | Special Operations Forces DAGRE Facility: Minor Construction | 7,200 | Heinrich |
| Air Force | New Mexico | Holloman Air Force Base | Holloman High-Speed Test Track: Design | 22,400 | Heinrich/Lujan |

# Congressionally Directed Spending Items (Continued)

| Account | State | Location | Project | Amount | Requestor(s) |
|---|---|---|---|---|---|
| Army | New Mexico | White Sands Missile Range | North Range Mission Control Center: Design | 3,330 | Heinrich/Lujan |
| Air Force Reserve | New York | Niagara Falls Air Reserve Station | TWY A4 and Runway 28R Overrun: Design | 5,400 | Gillibrand/Schumer |
| Army | New York | Fort Drum | Aircraft Maintenance Hangar Addition: Design | 9,300 | Gillibrand/Schumer |
| Army | New York | Fort Drum | Field Artillery Vehicle Storage Sheds: Design | 800 | Gillibrand/Schumer |
| Army | New York | Fort Drum | Fire Station 3 (WSAAF): Design | 2,760 | Gillibrand/Schumer |
| Army | New York | Fort Drum | ORTC Transient Training Barracks: Design | 7,790 | Gillibrand/Schumer |
| Air Force | North Carolina | Seymour Johnson Air Force Base | Combat Arms Training and Maintenance Complex | 15,000 | Budd |
| Air Force | Ohio | Wright-Patterson Air Force Base | Advanced Materials Research Laboratory—C2A: Design | 9,200 | Brown |
| Air Force | Ohio | Wright-Patterson Air Force Base | Dormitory: Design | 3,400 | Brown |
| Air Force | Ohio | Wright-Patterson Air Force Base | U.S. Space Force Intelligence Center: Design | 19,000 | Brown |
| Air Force | Oklahoma | Altus Air Force Base | Main Gate Entry Control Facility: Design | 2,200 | Mullin |
| Air National Guard | Oregon | Klamath Falls Airport—Kingsley Field | Aircraft Shelter MCCA: Design | 4,300 | Merkley/Wyden |
| Army National Guard | Oregon | Naval Weapons Systems Training Facility Boardman. | Multi-Purpose Machine Gun Range: Design | 1,690 | Merkley/Wyden |
| Defense-Wide (DLA) | Oregon | Klamath Falls Airport—Kingsley Field | Fuel Facility: Design | 3,000 | Merkley/Wyden |
| Air National Guard | Pennsylvania | Pittsburgh International Airport | Entry Control Facility: Minor Construction | 5,000 | Casey/Fetterman |
| Army National Guard | Pennsylvania | Fort Indiantown Gap Training Site | Ready Building: Design | 3,250 | Casey/Fetterman |
| Army Reserve | Pennsylvania | New Castle | Area Maintenance Support Activity and Land: Design | 1,750 | Casey/Fetterman |
| Army National Guard | Rhode Island | North Kingstown | National Guard Readiness Center: Cost to Complete | 16,000 | Reed |
| Air Force Reserve | South Carolina | Joint Base Charleston | Aeromedical Evacuation Facility | 33,000 | Graham |
| Navy and Marine Corps | South Carolina | Marine Corps Air Station Beaufort | Aircraft Maintenance Hangar: Cost to Complete | 61,700 | Graham |
| Army National Guard | Vermont | St. Albans | Readiness Center: Design | 4,030 | Welch |
| Navy and Marine Corps | Virginia | Marine Corps Base Quantico | Child Development Center: Design | 8,200 | Kaine/Warner |
| Navy and Marine Corps | Virginia | Naval Air Station Oceana | Unaccompanied Housing: Design | 16,000 | Kaine/Warner |
| Navy and Marine Corps | Virginia | Norfolk Naval Shipyard | Child Development Center: Design | 1,400 | Kaine/Warner |
| Air Force | Washington | Fairchild Air Force Base | KC–135 ADAL Aircraft Parts Warehouse: Design | 2,500 | Murray |
| Air Force | Washington | Fairchild Air Force Base | Taxilane Pull-Through Spots: Design | 4,500 | Murray |
| Air National Guard | Washington | Camp Murray | National Guard/Reserve Center (ANG Cost Share): Minor Construction. | 5,700 | Murray |
| Army | Washington | Yakima Training Center | Wastewater Treatment Plant: Design | 6,460 | Murray |
| Army National Guard | Washington | Tacoma | Controlled Humidity Warehouse: Minor Construction | 6,000 | Murray |
| Army National Guard | West Virginia | Bluefield | National Guard Readiness Center: Design | 4,000 | Capito/Manchin |
| Army National Guard | West Virginia | Camp Dawson | Fire Station Support Building: Minor Construction | 6,000 | Capito/Manchin |
| Army National Guard | West Virginia | Camp Dawson-Kingwood | General Instruction Building: Design | 1,200 | Capito/Manchin |
| Army National Guard | West Virginia | Joint Forces Headquarters—Charleston | National Guard/Reserve Center Building: Design | 1,400 | Capito/Manchin |
| Army National Guard | West Virginia | Joint Forces Headquarters—Charleston | Physical Fitness Center (Multipurpose Building): Design | 2,500 | Capito |
| Army National Guard | Wisconsin | Black River Falls | National Guard Readiness Center: Design | 2,600 | Baldwin |
| Army National Guard | Wisconsin | Wausau | National Guard Readiness Center: Design | 7,150 | Baldwin |
| Army National Guard | Wisconsin | Wisconsin Rapids | National Guard Readiness Center: Design | 4,940 | Baldwin |
| Army Reserve | Wisconsin | Manitowoc | Army Reserve Vehicle Maintenance Shop: Design | 150 | Baldwin |

## Comparative Statement of New Budget (Obligational) Authority for Fiscal Year 2024 and Budget Estimates and Amounts Recommended in the Bill for Fiscal Year 2025

[In thousands of dollars]

| Item | 2024 appropriation | Budget estimate | Committee recommendation | Senate Committee recommendation compared with (+ or −) | |
|---|---|---|---|---|---|
| | | | | 2024 appropriation | Budget estimate |
| TITLE I—DEPARTMENT OF DEFENSE | | | | | |
| Military Construction | | | | | |
| Active Components: | | | | | |
| Military Construction, Army | 2,022,775 | 2,311,157 | 2,380,477 | +357,702 | +69,320 |
| Military Construction, Navy and Marine Corps | 5,531,369 | 4,540,899 | 4,874,699 | −656,670 | +333,800 |
| Military Construction, Air Force | 2,741,424 | 3,187,126 | 3,549,626 | +808,202 | +362,500 |
| Military Construction, Defense-Wide | 3,161,782 | 3,733,163 | 3,601,163 | +439,381 | −132,000 |
| Subtotal, Active components | 13,457,350 | 13,772,345 | 14,405,965 | +948,615 | +633,620 |
| Reserve Components: | | | | | |
| Military Construction, Army National Guard | 620,647 | 362,129 | 456,459 | −164,188 | +94,330 |
| Military Construction, Air National Guard | 295,526 | 190,792 | 307,792 | +12,266 | +117,000 |
| Military Construction, Army Reserve | 151,076 | 255,032 | 375,682 | +224,606 | +120,650 |
| Military Construction, Navy Reserve | 51,291 | 29,829 | 29,829 | −21,462 | |
| Military Construction, Air Force Reserve | 331,572 | 69,263 | 107,663 | −223,909 | +38,400 |
| Subtotal, Reserve components | 1,450,112 | 907,045 | 1,277,425 | −172,687 | +370,380 |
| North Atlantic Treaty Organization Security Investment Program | 293,434 | 433,864 | 433,864 | +140,430 | |
| Department of Defense Base Closure Account | 489,174 | 447,961 | 497,961 | +8,787 | +50,000 |
| Total, Military Construction | 15,690,070 | 15,561,215 | 16,615,215 | +925,145 | +1,054,000 |
| Family Housing | | | | | |
| Family Housing Construction, Army | 304,895 | 276,647 | 276,647 | −28,248 | |
| Family Housing Operation and Maintenance, Army | 395,485 | 475,611 | 475,611 | +80,126 | |
| Family Housing Construction, Navy and Marine Corps | 277,142 | 245,742 | 245,742 | −31,400 | |
| Family Housing Operation and Maintenance, Navy and Marine Corps | 373,854 | 377,217 | 377,217 | +3,363 | |
| Family Housing Construction, Air Force | 237,097 | 221,549 | 221,549 | −15,548 | |

## Comparative Statement of New Budget (Obligational) Authority for Fiscal Year 2024 and Budget Estimates and Amounts Recommended in the Bill for Fiscal Year 2025 (Continued)

| | | | | | |
|---|---|---|---|---|---|
| Family Housing Operation and Maintenance, Air Force | 324,386 | 326,250 | 326,250 | +1,864 | |
| Family Housing Operation and Maintenance, Defense-Wide | 50,785 | 52,156 | 52,156 | +1,371 | |
| Department of Defense Family Housing Improvement Fund | 6,611 | 8,195 | 8,195 | +1,584 | |
| Department of Defense Military Unaccompanied Housing Improvement Fund | 496 | 497 | 497 | +1 | |
| Total, Family Housing | 1,970,751 | 1,983,864 | 1,983,864 | +13,113 | |
| Administrative Provisions | | | | | |
| Unfunded Requirements (Sec 124): | | | | | |
| Military Construction, Army | 8,214 | | 26,500 | +18,286 | +26,500 |
| Military Construction, Navy and Marine Corps | 182,150 | | 50,400 | −131,750 | +50,400 |
| Military Construction, Air Force | 166,300 | | 183,000 | +16,700 | +183,000 |
| Military Construction, Defense-Wide | 62,400 | | 134,270 | +71,870 | +134,270 |
| Military Construction, Army National Guard | 66,815 | | 54,251 | −12,564 | +54,251 |
| Military Construction, Air National Guard | 5,200 | | 6,000 | +800 | +6,000 |
| Military Construction, Army Reserve | 23,000 | | 63,000 | +40,000 | +63,000 |
| Military Installation Resilience (Sec 128): | | | | | |
| Military Installation Resilience, Army | 15,000 | | 10,000 | −5,000 | +10,000 |
| Military Installation Resilience, Navy and Marine Corps | 7,500 | | 10,000 | +2,500 | +10,000 |
| Military Installation Resilience, Air Force | 7,500 | | 10,000 | +2,500 | +10,000 |
| Foreign Military Training Infrastructure (Sec 129): | | | | | |
| Military Construction, Air Force | | | 100,000 | +100,000 | +100,000 |
| Military Construction, Defense-Wide | | | 2,500 | +2,500 | +2,500 |
| Military Construction, Air National Guard | 83,000 | | 26,000 | −57,000 | +26,000 |
| Authorized Projects (Sec 131) | | | | | |
| Military Construction, Air National Guard | | | 32,000 | +32,000 | +32,000 |
| Lab Infrastructure (Sec 128): | | | | | |
| Lab Infrastructure, Army | 10,000 | | | −10,000 | |
| Lab Infrastructure, Navy & Marine Corps | 10,000 | | | −10,000 | |
| Lab Infrastructure, Air Force | 10,000 | | | −10,000 | |
| Natural Disaster (Sec 129): | | | | | |
| Military Construction, Air Force | 150,000 | | | −150,000 | |
| Child Development Centers (Sec 130): | | | | | |
| Military Construction, Army | 15,000 | | | −15,000 | |
| Military Construction, Navy & Marine Corps | 15,000 | | | −15,000 | |
| Military Construction, Air Force | 15,000 | | | −15,000 | |

| Item | 2024 appropriation | Budget estimate | Committee recommendation | Senate Committee recommendation compared with (+ or −) | |
|---|---|---|---|---|---|
| | | | | 2024 appropriation | Budget estimate |
| Barracks Replacement (Sec 131): | | | | | |
| Military Construction, Army | 15,000 | | | − 15,000 | |
| Military Construction, Navy & Marine Corps | 15,000 | | | − 15,000 | |
| Military Construction, Air Force | 15,000 | | | − 15,000 | |
| Demolition (Sec 132): | | | | | |
| Military Construction, Army | 15,000 | | | − 15,000 | |
| Military Construction, Navy & Marine Corps | 15,000 | | | − 15,000 | |
| Military Construction, Air Force | 15,000 | | | − 15,000 | |
| Cost to Completes (Sec 133): | | | | | |
| Military Construction, Air Force | 16,000 | | | − 16,000 | |
| Family Housing Improvement (Sec 135): | | | | | |
| Family Housing, Navy & Marine Corps | 19,000 | | | − 19,000 | |
| Authorized Projects (Sec 136): | | | | | |
| Military Construction, Defense-Wide | 37,100 | | | − 37,100 | |
| Total, Administrative Provisions | 1,014,179 | | 707,921 | − 306,258 | + 707,921 |
| Total, title I, Department of Defense | 18,675,000 | 17,545,079 | 19,307,000 | + 632,000 | + 1,761,921 |
| Appropriations | (18,675,000) | (17,545,079) | (19,307,000) | (+ 632,000) | (+ 1,761,921) |
| TITLE II—DEPARTMENT OF VETERANS AFFAIRS | | | | | |
| Veterans Benefits Administration | | | | | |
| Compensation and pensions: | | | | | |
| Budget year appropriations | 15,072,388 | 9,820,699 | 9,820,699 | − 5,251,689 | |
| Advance from prior year appropriations | 146,778,136 | 182,310,515 | 182,310,515 | + 35,532,379 | |
| Subtotal, Compensation and pensions available in fiscal year | 161,850,524 | 192,131,214 | 192,131,214 | + 30,280,690 | |

## Comparative Statement of New Budget (Obligational) Authority for Fiscal Year 2024 and Budget Estimates and Amounts Recommended in the Bill for Fiscal Year 2025 (Continued)

| | | | | | |
|---|---|---|---|---|---|
| Advance appropriations, FY 2026 | 182,310,515 | 204,481,753 | 204,481,753 | +22,171,238 | |
| Less appropriations from prior year advances | −146,778,136 | −182,310,515 | −182,310,515 | −35,532,379 | |
| Total, Compensation and pensions appropriated in this bill | 197,382,903 | 214,302,452 | 214,302,452 | +16,919,549 | |
| Readjustment benefits: | | | | | |
| Budget year appropriations | 374,852 | 2,657,656 | 2,657,656 | +2,282,804 | |
| Advance from prior year appropriations | 8,452,500 | 13,399,805 | 13,399,805 | +4,947,305 | |
| Subtotal, readjustment benefits available in fiscal year | 8,827,352 | 16,057,461 | 16,057,461 | +7,230,109 | |
| Advance appropriations, FY 2026 | 13,399,805 | 17,614,235 | 17,614,235 | +4,214,430 | |
| Less appropriations from prior year advances | −8,452,500 | −13,399,805 | −13,399,805 | −4,947,305 | |
| Total, Readjustment benefits appropriated in this bill | 13,774,657 | 20,271,891 | 20,271,891 | +6,497,234 | |
| Veterans insurance and indemnities: | | | | | |
| Budget year appropriations | 12,701 | | | −12,701 | |
| Advance from prior year appropriations | 121,126 | 135,119 | 135,119 | +13,993 | |
| Subtotal, Veterans insurance and indemnities available in fiscal year | 133,827 | 135,119 | 135,119 | +1,292 | |
| Advance appropriations, FY 2026 | 135,119 | 131,518 | 131,518 | −3,601 | |
| Less appropriations from prior year advances | −121,126 | −135,119 | −135,119 | −13,993 | |
| Total, Veterans insurance and indemnities appropriated in this bill | 147,820 | 131,518 | 131,518 | −16,302 | |
| Veterans housing benefit program fund: | | | | | |
| Credit subsidy: | | | | | |
| Direct | −83,000 | 98,000 | 98,000 | +181,000 | |
| Guaranteed | 1,803,000 | 1,991,000 | 1,991,000 | +188,000 | |
| Subtotal (Credit subsidy) | 1,720,000 | 2,089,000 | 2,089,000 | +369,000 | |
| (Limitation on direct loans) | (500) | (500) | (500) | | |
| Administrative expenses | 316,742 | 319,596 | 319,596 | +2,854 | |
| Vocational rehabilitation loans program account | 78 | 64 | 64 | −14 | |
| (Limitation on direct loans) | (2,026) | (1,564) | (1,564) | (−462) | |
| Administrative expenses | 461 | 494 | 494 | +33 | |
| Native American veteran housing loan program account | 2,719 | | | −2,719 | |

| Item | 2024 appropriation | Budget estimate | Committee recommendation | Senate Committee recommendation compared with (+ or −) | |
|---|---|---|---|---|---|
| | | | | 2024 appropriation | Budget estimate |
| Administrative expenses | .................... | 5,845 | 5,845 | +5,845 | .................... |
| General operating expenses, VBA | 3,899,000 | 4,035,000 | 4,035,000 | +136,000 | .................... |
| Total, Veterans Benefits Administration | 217,244,380 | 241,155,860 | 241,155,860 | +23,911,480 | .................... |
| Budget year appropriations | (21,398,941) | (18,928,354) | (18,928,354) | (−2,470,587) | .................... |
| Advance appropriations, FY 2026 | (195,845,439) | (222,227,506) | (222,227,506) | (+26,382,067) | .................... |
| Advances from prior year appropriations | (155,351,762) | (195,845,439) | (195,845,439) | (+40,493,677) | .................... |
| Veterans Health Administration | | | | | |
| Medical services (MS): | | | | | |
| Budget year appropriations | −3,034,205 | .................... | .................... | +3,034,205 | .................... |
| Advance from prior year appropriations | 74,004,000 | 71,000,000 | 71,000,000 | −3,004,000 | .................... |
| Subtotal, MS available in fiscal year | 70,969,795 | 71,000,000 | 71,000,000 | +30,205 | .................... |
| Advance appropriations, FY 2026 | 71,000,000 | 75,039,000 | 75,039,000 | +4,039,000 | .................... |
| Total, MS appropriated in this bill | 67,965,795 | 75,039,000 | 75,039,000 | +7,073,205 | .................... |
| Medical community care (MCC): | | | | | |
| Budget year appropriations | −2,657,977 | .................... | .................... | +2,657,977 | .................... |
| Advance from prior year appropriations | 33,000,000 | 20,382,000 | 20,382,000 | −12,618,000 | .................... |
| Subtotal, MCC available in fiscal year | 30,342,023 | 20,382,000 | 20,382,000 | −9,960,023 | .................... |
| Advance appropriations, FY 2026 | 20,382,000 | 34,000,000 | 34,000,000 | +13,618,000 | .................... |
| Total, MCC appropriated in this bill | 17,724,023 | 34,000,000 | 34,000,000 | +16,275,977 | .................... |

# Comparative Statement of New Budget (Obligational) Authority for Fiscal Year 2024 and Budget Estimates and Amounts Recommended in the Bill for Fiscal Year 2025 (Continued)

| | | | | | |
|---|---|---|---|---|---|
| Medical support and compliance (MSC): | | | | | |
| Budget year appropriations | −1,550,000 | | | +1,550,000 | |
| Advance from prior year appropriations | 12,300,000 | 11,800,000 | 11,800,000 | −500,000 | |
| Subtotal, MSC available in fiscal year | 10,750,000 | 11,800,000 | 11,800,000 | +1,050,000 | |
| Advance appropriations, FY 2026 | 11,800,000 | 12,700,000 | 12,700,000 | +900,000 | |
| Total, MSC appropriated in this bill | 10,250,000 | 12,700,000 | 12,700,000 | +2,460,000 | |
| Medical facilities (MF): | | | | | |
| Budget year appropriations | 149,485 | | 3,000 | −146,485 | +3,000 |
| Advance from prior year appropriations | 8,800,000 | 9,400,000 | 9,400,000 | +600,000 | |
| Subtotal, MF available in fiscal year | 8,949,485 | 9,400,000 | 9,403,000 | +453,515 | +3,000 |
| Advance appropriations, FY 2026 | 9,400,000 | 9,700,000 | 9,700,000 | +300,000 | |
| Total, MF appropriated in this bill | 9,549,485 | 9,700,000 | 9,703,000 | +153,515 | +3,000 |
| Total, VA Medical Care available in fiscal year | 121,011,303 | 112,582,000 | 112,585,000 | −8,426,303 | +3,000 |
| Medical and prosthetic research | 943,000 | 868,000 | 878,000 | −65,000 | +10,000 |
| Medical care cost recovery collections: | | | | | |
| Offsetting collections | −3,991,000 | −4,632,000 | −4,632,000 | −641,000 | |
| Appropriations (indefinite) | 3,991,000 | 4,632,000 | 4,632,000 | +641,000 | |
| Subtotal | | | | | |
| Medical Care Collections Fund (MCCF) (transfer out) (Sec 214) | (−491,000) | (−501,000) | (−501,000) | (−10,000) | |
| Medical Community Care and Medical Services (by transfer) (Sec 214) | (491,000) | (501,000) | (501,000) | (+10,000) | |
| VA medical funds (transfer out) (Sec 219) | (−430,532) | (−594,828) | (−594,828) | (−164,296) | |
| DoD–VA Joint Medical Facility Demonstration Fund (by transfer) (Sec 219) | (430,532) | (594,828) | (594,828) | (+164,296) | |
| MCCF to DoD–VA Joint Medical Facility Demonstration Fund (transfer out) (Sec 221) | (−20,000) | (−17,000) | (−17,000) | (+3,000) | |
| DoD–VA Joint Medical Facility Demonstration Fund (by transfer) (Sec 221) | (20,000) | (17,000) | (17,000) | (−3,000) | |
| VA Medical Care Funds (transfer out) (Sec 222) | (−15,000) | (−15,000) | (−15,000) | | |
| DoD–VA Health Care Sharing Incentive Fund (by transfer) (Sec 222) | (15,000) | (15,000) | (15,000) | | |
| Total, Veterans Health Administration | 106,432,303 | 132,307,000 | 132,320,000 | +25,887,697 | +13,000 |

| Item | 2024 appropriation | Budget estimate | Committee recommendation | Senate Committee recommendation compared with (+ or −) | |
|---|---|---|---|---|---|
| | | | | 2024 appropriation | Budget estimate |
| Budget year appropriations | (−6,149,697) | (868,000) | (881,000) | (+7,030,697) | (+13,000) |
| Advance appropriations (FY 2026) | (112,582,000) | (131,439,000) | (131,439,000) | (+18,857,000) | |
| Advances from prior year appropriations | (128,104,000) | (112,582,000) | (112,582,000) | (−15,522,000) | |
| (By transfer) | (956,532) | (1,127,828) | (1,127,828) | (+171,296) | |
| (Transfer out) | (−956,532) | (−1,127,828) | (−1,127,828) | (−171,296) | |
| **National Cemetery Administration** | | | | | |
| National Cemetery Administration | 480,000 | 495,000 | 495,000 | +15,000 | |
| **Departmental Administration** | | | | | |
| General Administration | 475,000 | 457,000 | 457,000 | −18,000 | |
| Board of Veterans Appeals | 287,000 | 267,000 | 267,000 | −20,000 | |
| Information Technology Systems (IT Systems) | 6,401,000 | 6,231,680 | 6,276,693 | −124,307 | +45,013 |
| Veterans Electronic Health Record (EHR) | 1,334,142 | 894,000 | 894,000 | −440,142 | |
| Office of Inspector General | 296,000 | 296,000 | 301,000 | +5,000 | +5,000 |
| Construction, major projects | 961,219 | 2,069,000 | 2,069,000 | +1,107,781 | |
| Construction, minor projects | 692,000 | 380,453 | 380,453 | −311,547 | |
| Grants for construction of State extended care facilities | 171,000 | 141,000 | 150,000 | −21,000 | +9,000 |
| Grants for the construction of Veterans cemeteries | 60,000 | 60,000 | 65,000 | +5,000 | +5,000 |
| Total, Departmental Administration | 10,677,361 | 10,796,133 | 10,860,146 | +182,785 | +64,013 |
| **Cost of War Toxic Exposures Fund** | | | | | |
| Cost of War Toxic Exposures Fund, FY 26 Advance Appropriations | | 22,800,000 | | | −22,800,000 |
| **Administrative Provisions** | | | | | |
| Recurring Expenses Transformational Fund (Sec 252)(non-add) | (646,000) | | (307,000) | (−339,000) | (+307,000) |
| Veterans Electronic Health Record (Sec 256) (rescission) | −460,005 | | | +460,005 | |
| Air Ambulances discretionary (Sec 258) | | | 5,000 | +5,000 | +5,000 |
| Medical and Prosthetic Research (Sec 257) (rescission) | | | −5,000 | −5,000 | −5,000 |

## Comparative Statement of New Budget (Obligational) Authority for Fiscal Year 2024 and Budget Estimates and Amounts Recommended in the Bill for Fiscal Year 2025 (Continued)

| | | | | | |
|---|---|---|---|---|---|
| Air Ambulances discretionary (Sec 258) | ........ | ........ | ........ | ........ | ........ |
| Veterans Health Administration (Sec 259) (rescission) | −1,951,750 | ........ | ........ | +1,951,750 | ........ |
| Veterans Benefits Administration, General Operating Expenses (Sec 260) (rescission) | −30,000 | ........ | ........ | +30,000 | ........ |
| General Administration (Sec 260) (rescission) | −5,000 | ........ | ........ | +5,000 | ........ |
| Board of Veterans Appeals (Sec 260) (rescission) | −15,000 | ........ | ........ | +15,000 | ........ |
| Information Technology Systems (Sec 260) (rescission) | −15,000 | ........ | ........ | +15,000 | ........ |
| Major Construction (Sec 260) (rescission) | −80,219 | ........ | ........ | +80,219 | ........ |
| Bost Amendment (Chimp) | ........ | ........ | ........ | ........ | ........ |
| Bost Amendment (discretionary) | ........ | ........ | ........ | ........ | ........ |
| Total, Administrative Provisions | −2,556,974 | ........ | ........ | +2,556,974 | ........ |
| Total, title II, Department of Veterans Affairs | 332,277,070 | 407,553,993 | 384,831,006 | +52,553,936 | −22,722,987 |
| Budget year appropriations | (26,406,605) | (31,087,487) | (31,169,500) | (+4,762,895) | (+82,013) |
| Rescissions | (−2,556,974) | ........ | (−5,000) | (+2,551,974) | (−5,000) |
| Advance Appropriations, FY26: | | | | | |
| (Mandatory) | (195,845,439) | (245,027,506) | (222,227,506) | (+26,382,067) | (−22,800,000) |
| (Discretionary) | (112,582,000) | (131,439,000) | (131,439,000) | (+18,857,000) | ........ |
| (By transfer) | (956,532) | (1,127,828) | (1,127,828) | (+171,296) | ........ |
| (Transfer out) | (−956,532) | (−1,127,828) | (−1,127,828) | (−171,296) | ........ |
| (Limitation on direct loans) | (2,526) | (2,064) | (2,064) | (−462) | ........ |
| Advances from prior year appropriations: | | | | | |
| (Mandatory) | (155,351,762) | (195,845,439) | (195,845,439) | (+40,493,677) | ........ |
| (Discretionary) | (128,104,000) | (112,582,000) | (112,582,000) | (−15,522,000) | ........ |
| Total, available this fiscal year | 307,305,393 | 339,514,926 | 339,591,939 | +32,286,546 | +77,013 |
| (Mandatory) | (172,531,703) | (210,412,794) | (210,412,794) | (+37,881,091) | ........ |
| (Discretionary) | (134,773,690) | (129,102,132) | (129,179,145) | (−5,594,545) | (+77,013) |
| TITLE III—RELATED AGENCIES | | | | | |
| American Battle Monuments Commission | | | | | |
| Salaries and expenses | 158,630 | 89,520 | 94,520 | −64,110 | +5,000 |
| Total, American Battle Monuments Commission | 158,630 | 89,520 | 94,520 | −64,110 | +5,000 |

| Item | 2024 appropriation | Budget estimate | Committee recommendation | Senate Committee recommendation compared with (+ or −) | |
|---|---|---|---|---|---|
| | | | | 2024 appropriation | Budget estimate |
| US Court of Appeals for Veterans Claims | | | | | |
| Salaries and expenses | 47,200 | 47,300 | 47,300 | +100 | |
| Department of Defense—Civil | | | | | |
| Cemeterial Expenses, Army | | | | | |
| Salaries and expenses | 99,880 | 105,514 | 105,514 | +5,634 | |
| Construction | 88,600 | 42,000 | 42,000 | −46,600 | |
| Total, Cemeterial Expenses, Army | 188,480 | 147,514 | 147,514 | −40,966 | |
| Armed Forces Retirement Home—Trust Fund | | | | | |
| Operation and maintenance | 43,060 | 43,520 | 43,520 | +460 | |
| Capital program | 8,940 | 1,000 | 1,000 | −7,940 | |
| Payment from General Fund | 25,000 | 25,000 | 25,000 | | |
| Subtotal, Armed Forces Retirement Home—Trust Fund | 77,000 | 69,520 | 69,520 | −7,480 | |
| Armed Forces Retirement Home—Major Construction | | | | | |
| Major Construction | | 31,000 | 31,000 | +31,000 | |
| Total, Armed Forces Retirement Home | 77,000 | 100,520 | 100,520 | +23,520 | |
| Total, title III, Related Agencies | 471,310 | 384,854 | 389,854 | −81,456 | +5,000 |
| OTHER APPROPRIATIONS | | | | | |
| FISCAL RESPONSIBILITY ACT, 2023 (PL 118–5) | | | | | |
| Cost of War Toxic Exposures Fund, FY 2024 Budget Year (non-add) | (20,268,000) | | | (−20,268,000) | |

# Comparative Statement of New Budget (Obligational) Authority for Fiscal Year 2024 and Budget Estimates and Amounts Recommended in the Bill for Fiscal Year 2025 (Continued)

| | | | | | |
|---|---|---|---|---|---|
| Cost of War Toxic Exposures Fund, FY 2025 Advance Appropriation (non-add) | (24,455,000) | | | (−24,455,000) | |
| Total, Fiscal Responsibility Act, 2023 (PL 118–5) | 44,723,000 | | | −44,723,000 | |
| INDO—PACIFIC SECURITY SUPPLEMENTAL APPROPRIATIONS ACT, 2024 | | | | | |
| DEPARTMENT OF DEFENSE | | | | | |
| Military Construction, Navy and Marine Corps (emergency) | 281,917 | | | −281,917 | |
| Total, Indo-Pacific Security Supplemental | 281,917 | | | −281,917 | |
| Total, Other Appropriations | 281,917 | | | −281,917 | |
| Grand total | 351,705,297 | 425,483,926 | 404,527,860 | +52,822,563 | −20,956,066 |
| Budget year appropriations | (45,552,915) | (49,017,420) | (50,866,354) | (+5,313,439) | (+1,848,934) |
| Emergency appropriations in this bill | (281,917) | | | (−281,917) | |
| Rescissions | (−2,556,974) | | (−5,000) | (+2,551,974) | (−5,000) |
| Advance appropriations, FY26 | (308,427,439) | (376,466,506) | (353,666,506) | (+45,239,067) | (−22,800,000) |
| Advances from prior year appropriations | (283,455,762) | (308,427,439) | (308,427,439) | (+24,971,677) | |
| (By transfer) | (956,532) | (1,127,828) | (1,127,828) | (+171,296) | |
| (Transfer out) | (−956,532) | (−1,127,828) | (−1,127,828) | (−171,296) | |
| (Limitation on direct loans) | (2,526) | (2,064) | (2,064) | (−462) | |

## Chapter 3

# FY2025 NDAA: Military Construction and Housing Authorizations*

**Andrew Tilghman**

For fiscal year 2025 (FY2025), the Biden Administration requested $17.545 billion for U.S. Department of Defense (DOD) military construction (MILCON) accounts and family housing programs (a total that is 6% less than the $18.675 billion enacted by Congress for MILCON and family housing in the previous year).

On May 31, 2024, the House Armed Services Committee (HASC) reported the Servicemember Quality of Life Improvement and National Defense Authorization Act (NDAA) for Fiscal Year 2025 (H.R. 8070), and an accompanying report, H.Rept. 118-529. On June 14, the House passed H.R. 8070, which would authorize $17.545 billion for DOD MILCON and family housing accounts, equal to the Administration's requested amount.

On July 8, 2024, the Senate Armed Services Committee (SASC) reported S. 4638, the National Defense Authorization Act for Fiscal Year 2025, and an accompanying report, S.Rept. 118-188. The legislation would authorize $20.664 billion for traditional military construction and family housing accounts, 15% more than the requested amount. The SASC-reported bill would also provide an additional $12.5 billion to repair Air Force and Navy facilities on Guam that were damaged by a 2023 typhoon. The additional disaster relief funding—which is not present in H.R. 8070—brings S. 4638's total MILCON and family housing authorization to $33.169 billion, 89% more than the Administration's requested amount.

---

* This is an edited, reformatted and augmented version of Congressional Research Service Publication No. IN12411, dated August 28, 2024.

In: Military Construction and Appropriations
Editor: George K. Santiago
ISBN: 979-8-89530-799-1

Details of the funding levels for selected MILCON and family housing accounts are contained in Table 1.

**Table 1.** FY2025 NDAA authorizations for selected military construction accounts

(in thousands of dollars of discretionary budget authority)

| | *FY2024 Enacted* | *FY25 President's Budget Request* | *H.R. 8070* | *S. 4638* |
|---|---|---|---|---|
| Military Construction, Army | $2,022,775 | $2,311,157 | $2,149,957 | $2,361,328 |
| Military Construction, Navy | $5,531,369 | $4,540,899 | $4,104,429 | $6,547,616 |
| Military Construction, Air Force | $2,741,424 | $3,187,126 | $3,410,837 | $3,568,766 |
| Military Construction, Defense-wide | $3,161,782 | $3,733,163 | $3,636,722 | $3,735,946 |
| Military Construction, Army National Guard | $620,647 | $362,129 | $448,529 | $555,181 |
| Military Construction, Air National Guard | $295,526 | $190,792 | $238,792 | $314,192 |
| Military Construction, Army Reserve | $151,076 | $255,032 | $429,032 | $352,632 |
| Military Construction, Navy and Marine Reserve | $51,291 | $29,829 | $114,829 | $104,829 |
| Military Construction, Air Force Reserve | $331,572 | $69,263 | $121,263 | $96,283 |
| NATO Security Investment Program | $293,434 | $433,864 | $433,864 | $463,864 |
| Family Housinga | $1,970,751 | $1,983,864 | $1,933,864 | $1,965,864 |
| Base Realignment and Closure (BRAC) | $489,174 | $447,961 | $522,961 | $447,961 |
| INDOPACOM MILCON Pilot | $0 | $0 | $0 | $150,000 |
| Total Military Construction, Family Housing and BRAC | $18,675,000 | $17,545,079 | $17,545,079 | $20,664,462 |
| Disaster Recovery, Navy (Guam) | $0 | $1,224,771[b] | $0 | $4,566,940 |
| Disaster Recovery, Air Force (Guam) | $0 | $71,197[b] | $0 | $7,938,000 |
| *Total Division B, Military Construction* | *$18,675,000* | *$18,841,047* | *$17,545,079* | *$33,169,402* |

Source: P.L. 118-42; President's Budget Request FY2025, Office of the Under Secretary of Defense (Comptroller), Construction Programs (C-1), March 2024; White House Letter Regarding Critical Supplemental Funding Needs, June 2024, and Funding Request to Meet Critical Needs, H.R. 8070, Sec. 4601, and S. 4368, Sec. 4601 and 4602.

Notes: [a] Family Housing budget numbers include Family Housing Construction accounts, Family Housing Operation and Maintenance accounts, the Family Housing Improvement Fund (FHIF), and the Unaccompanied Housing Improvement Fund (UHIF).

[b] The Biden Administration's request for disaster recovery was not included in the President's Budget Request submitted to Congress in March 2024, but was submitted under a separate supplemental request, first in October 2023 and resubmitted in June 2024.

## H.R. 8070

The House bill contains 36 Community Project Funding items, totaling $1.338 billion.

The House bill contains certain provisions that may impact laws, funding, and oversight mechanisms related to military construction and family housing programs. Selected provisions from the bill are listed and briefly described below.

Section 2801 would authorize the development and operation of a "Naval Innovation Center" at the Naval Postgraduate School in Monterey, California; the provision would also authorize the Secretary of Defense to accept financial gifts to support the center.

Section 2802 would authorize DOD's Office of Local Defense Community Cooperation (OLDCC) to expand a grant program that supports local initiatives for economic adjustment and diversification.

Specifically, the provision would authorize grants to local governments (in addition to state governments), along with grants for "public infrastructure projects and services that enhance the capabilities and resilience of the defense industrial base and the defense industrial base workers."

Section 2803 would authorize DOD's OLDCC to expand certain grant programs to include the Coast Guard.

Section 2804 would authorize DOD's OLDCC to expand the Defense Community Infrastructure Program (DCIP) to include "not-for-profit, member-owned utility services" among eligible recipients.

Section 2805 would raise spending limits for certain aspects of the Defense Laboratory Modernization Program.

Section 2807 would expand authority for funding MILCON projects for stormwater management and defense access roads.

Section 2808 would amend the FY2024 NDAA provision that increased the threshold for unspecified minor military construction (UMMC) projects in the Indo-Pacific Command's (INDOPACCOM's) Area of Responsibility. Specifically, the provision in Section 2808 would raise the threshold for UMMC from $15 million to $20 million.

Section 2822 would impose additional requirements for the development of a DOD-wide database of complaints related to military housing. Similarly, Section 2825 would restrict availability of certain DOD funds until DOD implements existing law requiring creation of a database of military housing complaints.

## S. 4638

The SASC-reported bill contains certain provisions that may impact laws, funding, and oversight mechanisms related to military construction and family housing programs. Selected provisions from the bill are listed and briefly described below.

Section 2801 would amend the statute governing UMMC (10 U.S.C. §2805) to authorize the Secretary of Defense to acquire land with UMMC appropriations.

Section 2802 would extend temporary authority for the Secretary of Defense to convey property at military installations.

Section 2804 would amend the FY2024 NDAA provision that increased the threshold for UMMC projects in the INDOPACCOM's Area of Responsibility. Specifically, the provision would raise the threshold for minor military construction from $15 million to $30 million.

Section 2805 would amend the statute (10 U.S.C. §2854) authorizing the restoration or replacement of damaged or destroyed facilities. The provision would require that any restoration or replacement project be designed to prevent damage or destruction by the cause or causes that generated the original damage or destruction.

Section 2806 would expand the law that authorizes DOD to lease certain non-excess property;

specifically, the provision would add additional types of in-kind consideration that DOD may accept as a part of such leasing.

Section 2808 would impose new requirements on DOD for obligating and executing funds appropriated for the purpose of designing a military construction project.

Section 2810 would impose a new requirement to limit the aggregate growth of square footage of DOD buildings. Specifically, the provision requires that for any square footage of growth in a military building, DOD must offset that amount of growth in space by an equivalent reduction of square footage in another facility by means such as demolition or disposal.

Section 2811 would create a new pilot program intended to optimize the Army's infrastructure footprint by authorizing the use of UMMC funding for certain projects that the Secretary of the Army determines will reduce overall square footage and "create more livable, productive, and resilient communities."

Section 2015 would set new requirements for the military departments to provide funding for facilities sustainment, restoration and modernization

(FSRM). A subcomponent of the services' Operation and Maintenance accounts, FSRM funding is intended to improve the condition of existing infrastructure.

Specifically, the provision in Section 2015 would require the military departments to invest FSRM funds in an amount equal to certain percentages of the total plant replacement value (PRV) of all facilities maintained by the department. The required investment would begin in FY2026 at 1.75% of total PRV. The requirement would increase incrementally until FY2029, when the requirement would reach 4% of total PRV. The legislation does not indicate any requirements for FSRM investments beyond FY2029.

Chapter 4

# FY2025 Military Construction Appropriations: A Summary*

**Andrew Tilghman**

The Biden Administration's fiscal year 2025 (FY2025) budget submission to Congress requested $17.545 billion for U.S. Department of Defense (DOD) military construction (MILCON) and family housing programs. The funding requested for FY2025 was 6.1% percent less than the $18.675 billion enacted for MILCON and family housing in FY2024 in Division A of P.L. 118-42.

On May 28, 2024, the House Appropriations Committee reported a Military Construction, Veterans Affairs, and Related Agencies Appropriation Act, 2025, H.R. 8580 (H.Rept. 118-528). On June 5, 2024, the House passed H.R. 8580, which would provide $17.957 billion for DOD MILCON accounts, 2.3% more than the requested amount.

On July 11, 2024, the Senate Appropriations Committee reported a version of the bill, S. 4677 (S.Rept. 118-191). The reported bill would provide $19.307 billion for DOD MILCON and family housing accounts, 10.0% more than the requested amount.

Details of the funding levels for selected MILCON and family housing accounts are contained in Table 1.

---

* This is an edited, reformatted and augmented version of Congressional Research Service Publication No. IN12407, dated August 20, 2024.

In: Military Construction and Appropriations
Editor: George K. Santiago
ISBN: 979-8-89530-799-1

**Table 1.** FY2024 MILCON-VA appropriations for selected military construction accounts

(in thousands of dollars of discretionary budget authority)

| *Account* | *FY2024 Enacted* | *FY2025 President's Budget Request* | *House-passed* | *Senate Appropriations Committee-reported* |
|---|---|---|---|---|
| Military Construction, Army | $2,022,775 | $2,311,157 | $2,217,757 | $2,380,477 |
| Military Construction, Navy | $5,531,369 | $4,540,899 | $4,332,414 | 4,874,699 |
| Military Construction, Air Force | $2,741,424 | $3,187,126 | $3,268,276 | $3,549,626 |
| Military Construction, Defense-wide | $3,151,782 | $3,733,163 | $3,500,083 | $3,601,163 |
| Military Construction, Army National Guard | $620,647 | $362,129 | $367,129 | $456,459 |
| Military Construction, Air National Guard | $295,526 | $190,792 | $195,792 | $307,792 |
| Military Construction, Army Reserve | $151,076 | $255,032 | $265,032 | $375,682 |
| Military Construction, Navy and Marine Reserve | $51,291 | $29,829 | $67,329 | $29,829 |
| Military Construction, Air Force Reserve | $331,572 | $69,263 | $50,499 | $107,663 |
| NATO Security Investment Program | $293,434 | $433,864 | $433,864 | $433,864 |
| Family Housing | $1,970,751 | $1,983,864 | $2,013,864 | $1,983,864 |
| Base Realignment and Closure | $489,174 | $447,961 | $547,961 | $497,961 |
| Administrative Provisions | $1,014,179 | $0 | $697,000 | $707,921 |
| *Total Division B, Military Construction* | *$18,675,000* | *$17,545,079* | *$17,957,000* | *$19,307,000* |

Source: P.L. 118-42; President's Budget Request FY2025, Office of the Under Secretary of Defense (Comptroller), Construction Programs (C-1), March 2024; H.R. 8580; and S. 4677.

Notes: Family Housing budget numbers include Family Housing Construction accounts, Family Housing Operation and Maintenance accounts, the Family Housing Improvement Fund (FHIF), and the Unaccompanied Housing Improvement Fund (UHIF).

## President's Budget Request

The Biden Administration's FY2025 budget requested funding for 150 major military construction projects. The 10 biggest projects, in terms of dollar value, are listed in *Table 2*.

**Table 2.** Ten Largest MILCON projects in the fy2025 president's budget request

(in thousands of dollars of budget authority)

| *Location* | *Project* | *Amount requested* |
|---|---|---|
| Joint Base Pearl Harbor-Hickam (HI) | Dry Dock 3 Replacement (Inc) | $1,199,000 |
| Portsmouth Naval Shipyard (ME) | Multi-Mission Drydock #1 Extension (Inc) | $400,578 |
| Joint Region Marianas | PDI: GDS, EIAMD, Ph1 (Inc) | $278,267 |
| Fort Meade (MD) | NSAW East Campus Building #5, INC 2 | $265,000 |
| Malmstrom Air Force Base (MT) | Weapons Storage & Maintenance Fac Inc | $238,000 |
| Wheeler Army Air Field (HI) | Aircraft Maintenance Hangar | $231,000 |
| Fort Belvoir (VA) | Defense Health Headquarters | $225,000 |
| Cape Canaveral Space Force Station (FL) | Engineering Test Facility | $221,060 |
| Cherry Point Marine Corps Air Station (NC) | Aircraft Maintenance Hangar | $213,520 |
| Marine Corps Base Hawaii (HI) | Aircraft Hangar & Parking Apron | $203,520 |

Source: Construction Programs (C-1), Department of Defense Budget Fiscal Year 2025 March 2024 Office of the Under Secretary of Defense (Comptroller).

Notes: The table shows the 10 largest MILCON projects in the President's Budget Request, in terms of dollar value. The "(Inc)" indicator listed with the title or description of some projects indicates that DOD has designated the project as an "incremental" part of a larger, multi-year MILCON project.

## H.R. 8580

H.R. 8580, as passed by the House, would revise funding levels for certain MILCON projects requested by the Administration. Table 3 lists the largest proposed adjustments in terms of dollar value. For example, the revisions include providing half of the amounts DOD requested for an aircraft maintenance hangar for Wheeler Army Air Field in Hawaii and an aircraft maintenance hangar for Marine Corps Air Station Cherry Point in North Carolina. In both instances, the reductions are examples of incremental funding, i.e., when Congress provides a portion of the requested amount for a particular project. By

incrementally funding certain projects, Congress allows DOD to initiate the construction project but would likely oblige DOD to request additional funding in a future budget cycle to complete the project. For the projects at Wheeler Army Air Field and at Marine Corps Air Station Cherry Point, DOD budget request documents included planned construction schedules indicating that construction of the project would take several years and DOD intends to outlay only an incremental portion of the project's funding during FY2025. The construction schedules indicate that DOD does not foresee a need to outlay the entire portion of the projects' costs until FY2026 or beyond.

The House-passed bill would provide incremental funding for 11 projects. The Biden Administration opposed incremental funding of MILCON projects. The House report (H.Rept. 118-528) accompanying the legislation states: "In general, the Committee supports full funding for military construction projects if they are executable. However, it continues to be the practice of the Committee to provide incremental funding for certain large projects to enable the Services to more efficiently allocate military construction dollars among projects that can be executed in the year of appropriation."

H.R. 8580 as passed by the House would provide funding for 23 unrequested Community Project Funding items, totaling $547 million.

Administrative provisions in H.R. 8580 would provide additional funding for planning and design of DOD laboratories, Air Force expenses incurred as a result of natural disasters, planning and design of child development centers, and planning and design of barracks.

**Table 3.** Proposed funding for selected military construction projects in H.R. 8580

(in thousands of dollars of budget authority)

| *Location* | *Project Title* | *Requested Amount* | *H.R. 8580* |
|---|---|---|---|
| Wheeler Army Air Field (HI) | Aircraft Maintenance Hangar | $231,000 | $115,500 |
| Marine Corps Air Station Cherry Point (NC) | Aircraft Maintenance Hangar | $213,520 | $106,760 |
| Naval Base Kitsap (WA) | Launcher Equipment Processing Building | $200,550 | $100,275 |
| Cape Canaveral SFS (FL) | Engineering Test Facility | $221,060 | $139,130 |
| Redstone Arsenal (AL) | Ground Test Facility | $80000 | $0 |
| Camp Butler (Japan) | Kubasaki High School | $160,000 | $80,000 |
| RAF Lakenheath (UK) | Lakenheath High School | $153,000 | $73,000 |

| *Location* | *Project Title* | *Requested Amount* | *H.R. 8580* |
|---|---|---|---|
| F.E. Warren (WY) | GBSD Consolidated Maintenance Facility | $194,000 | $115,000 |
| Lackland Air Force Base (TX) | Military Training Classrooms/Dining Facility | $0 | $70,000 |
| Ebbing Air National Guard Base (AR) | Academic Training Center | $0 | $70,000 |
| Naval Air Station Key West (FL) | Joint Interagency Task Force-South Command and Control Facility | $0 | $70,000 |

Source: CRS analysis of H.R. 8580 and Construction Programs (C-1), Department of Defense Budget Fiscal Year 2025 March 2024 Office of the Under Secretary of Defense (Comptroller). A list of project line items and the changes made in H.R. 8580 are found in the Appropriations Committee's report, H.Rept. 118-528, beginning on p. 88.

Notes: The table shows the 11 MILCON projects in H.R. 8580 with the largest proposed difference in funding compared to the President's Budget request. The "(Inc)" indicator listed with the title or description of some projects denotes the DOD has designated the project as an "incremental" part of a larger, multi-year MILCON project.

## S. 4677

S. 4677, as reported by the Senate Appropriations Committee, would revise funding levels for certain MILCON projects requested in the President's budget. *Table 4* lists the largest proposed adjustments in terms of dollar value. The legislation includes funding for projects that DOD did not include in its request, for example two projects for water treatment plants at Marine Corps Air Station Yuma in Arizona and Joint Base Pearl Harbor in Hawaii. In both instances, the projects were specifically requested for inclusion in the bill by the Senators representing those states.

Administrative provisions in the bill would provide additional funding for military installation resilience projects, as well as design and authorized construction of projects at foreign military training sites.

The Committee-reported S. 4677 recommends funding for 106 unrequested Congressionally Directed Spending items, totaling $1.264 billion.

**Table 4.** Proposed funding for selected military construction projects in S. 4677

(in thousands of dollars of budget authority)

| Location | Project Title | Requested Amount | S. 1677 |
|---|---|---|---|
| Fort Johnson (LA) | Barracks | $117,000 | $0 |

**Table 4.** (Continued)

| Location | Project Title | Requested Amount | S. 1677 |
|---|---|---|---|
| Marine Corps Air Station Yuma (AZ) | Water Treatment Plant | $0 | $90,000 |
| Joint Base Pearl Harbor-Hickam (HI) | Water Treatment Plant | $0 | $90,000 |
| Redstone Arsenal (AL) | Ground Test Facility Infrastructure (INC) | $80,000 | 0 |
| Fort Knox (KY) | Aviation Support Facility | $0 | $75,000 |
| Offutt Air Force Base (NE) | Survivable Airborne Operations Center: Design | $0 | $70,000 |
| Marine Corps Air Station Beaufort (SC) | Aircraft Maintenance Hangar: Cost to Complete | $0 | $61,700 |
| Buckley Space Force Base (CO) | Power Independence | $0 | $60,000 |
| Naval Base Coronado (CA) | SOF Operations Support Facility PH 2 | $51,000 | 0 |
| Naval Surface Warfare Center Indian Head (MD) | Contained Burn Facility | $0 | $50,000 |
| Bangor International Airport (ME) | Fuel Cell Hangar | $0 | $50,000 |

Source: CRS analysis of S. 4677 and Construction Programs (C-1), Department of Defense Budget Fiscal Year 2024, March 2023, Office of the Under Secretary of Defense (Comptroller). A list of project line items and the changes made in S. 4677 are found in the Appropriations Committee's report, S.Rept. 118-191 p 79.

Notes: The table shows the 11 MILCON projects in S. 4677 with the largest proposed change in funding compared to the President's Budget request. The "(Inc)" indicator listed with the title or description of some projects denotes the project as an "incremental" part of a larger, potentially multi-year MILCON project.

Chapter 5

# FY2024 NDAA: Military Construction and Housing Authorizations*

**Andrew Tilghman**

The Biden Administration's Fiscal Year 2024 (FY2024) budget requested $16.675 billion for U.S. Department of Defense (DOD) Military Construction (MILCON) accounts, which included funding for Military Construction and Family Housing programs. The funding requested was 12.4% percent less than the FY2023 enacted amount.

The House Armed Services Committee-reported version of an FY2024 NDAA (H.R. 2670; H.Rept. 118- 125) would have authorized $17.475 billion for DOD's MILCON programs—$800 million (4.8%) more than requested. Of that amount, H.R. 2670 would have authorized 26 earmarks, known as Community Project Funding items, totaling $480 million.

The Senate Armed Services Committee-reported FY2024 NDAA (S. 2226; S.Rept. 118-58) would have authorized $16.675 billion for DOD's MILCON programs, as requested. S. 2226 did not contain any earmarks, known as Congressionally Directed Spending items.

Congress passed the FY2024 NDAA (P.L. 118-31; H.Rept. 118-301), which became law on December 22, 2023. The law authorized $18.175 billion for Military Construction (MILCON) accounts—$1.5 billion more (9.0%) more than the Biden Administration's request. Of that amount, the law contained 25 Community Project Funding items totaling $559.4 million. See Table 1.

* This is an edited, reformatted and augmented version of Congressional Research Service Publication No. IN12229, Updated January 17, 2024.

In: Military Construction and Appropriations
Editor: George K. Santiago
ISBN: 979-8-89530-799-1

**Table 1.** FY2024 NDAA funding authorizations for selected military construction accounts

(in thousands of dollars of discretionary budget authority)

| *Account* | *FY2024 Request* | *House Authorized* | *Senate Authorized* | *Enacted* |
|---|---|---|---|---|
| Military Construction, Army | $1,470,555 | $1,803,165 | 1,644,779 | $1,912,289 |
| Military Construction, Navy | $6,022,187 | $5,412,514 | 4,668,487 | $5,304,040 |
| Military Construction, Air Force | $2,605,314 | $3,045,314 | 3,071,814 | $3,158,014 |
| Military Construction, Defense-wide | $2,984,682 | $2,856,147 | $3,162,706 | $3,198,571 |
| Military Construction, Army National Guard | $340,186 | $590,942 | $650,568 | $732,078 |
| Military Construction, Army Reserve | $107,076 | $225,076 | $170,076 | $234,076 |
| Military Construction, Navy and Marine Reserve | $51,291 | $66,291 | $51,291 | $66,291 |
| Military Construction, Air National Guard | $178,722 | $331,064 | $322,292 | $361,265 |
| Military Construction, Air Force Reserve | $291,572 | $344,072 | $309,572 | $346,572 |
| NATO Security Investment Program | $293,434 | $293,434 | $293,434 | $343,434 |
| Family Housing[a] | $1,933,644 | $1,960,644 | $1,933,644 | $1,972,033 |
| Family Housing Improvement Fund (FHIF) | $6,611 | $6,611 | $6,611 | $6,611 |
| Unaccompanied Housing Improvement Fund (UHIF) | $496 | $496 | $496 | $496 |
| Base Realignment and Closure[b] | $389,174 | $539,174 | $389,174 | $539,174 |
| Total Division B, Military Construction | $16,674,944 | $17,474,944 | $16,674,944 | $18,174,944 |

Source: U.S. Congress, Conference Committee, National Defense Authorization Act for Fiscal Year 2024, conference report to accompany H.R. 2670, 118th Cong., 1st sess., H.Rept. 118-301, December 6, 2023, pp. 1391-1392. For House and Senate authorized budget numbers, see funding tables in H.Rept. 118-125, Sec. 4601; and S.Rept. 118-58, Sec. 4601.

Notes:

[a] Family Housing amounts reflect a subtotal of amounts for the following accounts: Family Housing Construction, Army; Family Housing Operation & Maintenance, Army; Family Housing Construction, Navy and Marine Corps; Family Housing Operation & Maintenance, Navy and Marine Corps; Family Housing Construction, Air Force; Family Housing Operation & Maintenance, Air Force; Family Housing Operation & Maintenance, Defense-Wide.

[b] Base Realignment and Closure amounts reflect a subtotal of amounts for the following accounts: Base Realignment and Closure—Army, Base Realignment and Closure—Navy; Base Realignment and Closure—Air Force; and Base Realignment and Closure—Defense-Wide.

## Selected Policy Provisions

Policy provisions in the FY2024 NDAA that may affect military construction and housing programs include the following:

### Military Construction

Section 2801 amended 10 U.S.C. §2391 to make the Defense Community Infrastructure Pilot (DCIP) program permanent.

Section 2802 modified thresholds for minor MILCON as codified in 10 U.S.C. §2805 to define an unspecified minor MILCON project as a construction project that has an approved cost equal to or less than $9 million.

Section 2805 created 10 U.S.C. §2817, expanding authority to use O&M funds for certain MILCON projects in friendly foreign countries if the projects support air or sea ports of debarkation, or rail or other logistical support locations.

Section 2807 granted authority to utilize cost-plus incentive-fee contracting for MILCON projects associated with the Sentinel intercontinental ballistic missile weapons system.

Section 2810 raised the threshold for minor MILCON projects in the Indo-Pacific Command (INDOPACOM) area of operations to define an unspecified minor MILCON project as one that has an approved cost up to $15 million. The law also authorized $62 million for unspecified minor MILCON in INDOPACOM.

### Privatized Housing

Section 2821 establishes a DOD Military Family Readiness Working Group for Military Housing.

Section 2822 amended 10 U.S.C. §2878(f)(2), requiring DOD to notify Congress when negotiating with a privatized housing company for the extension of any ground lease of property. The provision requires notification about whether the privatized housing company is in compliance with the Military Housing Privatization Initiative (MHPI) Tenant Bill of Rights codified at 10 U.S.C. §2890. The provision also prohibits certain payments to

MHPI companies unless DOD determines that the company is in compliance with the MHPI Tenant Bill of Rights.

Section 2823 requires MHPI companies to notify tenants about a proposed nondisclosure agreement and provide a waiting period of 10 business days so tenants may seek legal counsel before being required to sign such an agreement.

Section 2825 requires the Secretary of Defense to implement the recommendations made in an April 2023 report from the U.S. Government Accountability Office.

## Unaccompanied Housing

Section 2831 amended 10 U.S.C. §2856 to require the Secretary of Defense to establish and enforce standards for DOD unaccompanied housing facilities regarding design, floor space, and level of maintenance required. The provision requires that all housing assigned to servicemembers meet the new requirements within two years of the law's enactment.

Section 2832 required the Secretary of Defense to establish standards for unaccompanied housing facilities regarding condition; safety and security; and "habitability, health and environmental comfort."

Section 2833 modified the procedures for a military department to waive requirements related to unaccompanied housing standards. The provision created 10 U.S.C. 2856a, requiring the Secretary of a military department to approve any such waivers and stipulates that the Secretary may do so only after exhausting options that include a) use of available privately-owned military housing; b) modifying the "unit integrity" policies that typically require servicemembers from the same unit to live in the same housing facility; and c) authorizing servicemembers to receive a Basic Allowance for Housing (BAH).

Section 2834 required the Secretary of Defense to submit annual certifications that the repairs and improvements needed for unaccompanied housing facilities do not exceed 20% of the total replacement costs of such facilities, a standard set by current DOD policy.

Section 2835 authorized a pilot program for replacing substandard enlisted barracks using O&M or unspecified minor MILCON funding.

Section 2836 required the establishment a civilian employee at the housing office at each military installation to oversee unaccompanied housing facilities and related issues.

Section 2837 requires the Secretary of Defense to issue rules for managing work orders related to maintenance work for unaccompanied housing facilities and to establish performance metrics to track the maintenance work performed.

Section 2838 requires the establishment of a DOD-wide index for evaluating the condition of unaccompanied housing facilities.

Section 2839 requires annual reports to Congress on the condition of unaccompanied military housing facilities based on the uniform index required under Section 2838.

Section 2840 requires the Secretary of Defense to notify Members of Congress if the Secretary enters into any agreement that provides DOD housing facilities for use by another federal agency in the Member's district.

Section 2841 required DOD to eliminate the use of waivers for construction standards for new unaccompanied housing facilities.

**Installation Management**

Section 2851 required the Secretary of Defense to initiate standardization of rules for installation access across DOD.

Chapter 6

# FY2024 Military Construction Appropriations: A Summary*

**Andrew Tilghman**

On March 9, 2023, the Biden Administration submitted the FY2024 President's Budget Request (PBR) seeking $16.675 billion for U.S. Department of Defense (DOD) military construction (MILCON) and family housing programs. The request was $2.325 billion (12%) less than the $19.000 billion provided for such programs in the Military Construction, Veterans Affairs and Related Agencies Appropriations Act, 2023 (Division J of P.L. 117-328). The House-passed version of the legislation for FY2024 (H.R. 4366) proposed appropriating $799 million (4.8%) more than the requested amount. The Senate Appropriations Committee-reported version of the legislation (S. 2127) recommended $2.395 billion (14.4%) more than the requested amount. On March 9, 2024, President Biden signed P.L. 118-42, which provided $18.675 billion for the FY2024 MILCON and family housing programs, 12% above the requested amount (Table 1). Explanatory materials and bill funding tables for Division A, the Military Construction, Veterans Affairs and Related Agencies Appropriations (MILCON-VA) Act are contained in the March 5, 2024 Congressional Record.

A list of selected projects funded in the legislation is included in Table 2.

The FY2024 MILCON-VA appropriations act included 11 projects that were incrementally funded (i.e., Congress provided a portion of the requested amount for a particular project). By incrementally funding certain projects, Congress allows DOD to initiate the construction project but DOD will likely have to request additional funding in a future budget cycle to complete the

* This is an edited, reformatted and augmented version of Congressional Research Service Publication No. IN12238, Updated April 25, 2024.

In: Military Construction and Appropriations
Editor: George K. Santiago
ISBN: 979-8-89530-799-1

project. In July 2023, the Biden Administration cited opposition to the House-passed version of the MILCON-VA bill because it proposed funding 24 projects incrementally.

The act appropriated funding for 143 MILCON projects that were not included in the President's budget request but were listed as Community Project Funding or Congressionally Directed Spending items.

Appropriations for those projects totaled about $1.565 billion.

**Table 1.** Military construction appropriation legislation for FY2024

(in thousands of dollars of budget authority)

| *Account* | *FY2023 Enacted* | *FY2024 Request* | *H.R. 4366* | *S. 2127* | *FY2024 Enacted* |
|---|---|---|---|---|---|
| Military Construction, Army | $1,553,825 | $1,470,555 | $1,517,455 | $1,876,875 | $2,022,775 |
| Military Construction, Navy & Marine | $4,345,320 | $6,022,187 | $4,477,961 | $6,046,309 | $5,531,369 |
| Military Construction, Air Force | $2,614,996 | $2,605,314 | $2,439,614 | $2,802,942 | $2,741,424 |
| Military Construction, Defense-wide | $2,626,078 | $2,984,682 | $2,651,047 | $3,132,782 | $3,161,782 |
| Military Construction, Army National Guard | $459,018 | $340,186 | $369,261 | $598,572 | $620,647 |
| Military Construction, Army Reserve | $193,878 | $107,076 | $117,076 | $151,076 | $151,076 |
| Military Construction, Navy & Marine Reserve | $36,837 | $51,291 | $61,291 | $51,291 | $51,291 |
| Military Construction, Air National Guard | $279,353 | $178,722 | $189,322 | $304,426 | $295,526 |
| Military Construction, Air Force Reserve | $85,423 | $291,572 | $330,572 | $309,572 | $331,572 |
| NATO Security Investment Program | $220,139 | $293,434 | $293,434 | $293,434 | $293,434 |
| Base Realignment and Closure | $574,687 | $389,174 | $539,174 | $439,174 | $489,174 |
| *Subtotal, Military Construction* | *$12,989,554* | *$14,734,193* | *$12,986,207* | *$16,006,435* | *$15,690,070* |
| *Subtotal, Family Housing* | *$1,986,330* | *$1,940,751* | *$1,970,751* | *$1,940,751* | *$1,970,751* |
| *Subtotal, Administrative Provisions* | *$4,024,116* | *$0* | *$2,517,042* | *$1,122,814* | *$1,014,179* |
| *Total, Military Construction* | *$19,000,000* | *$16,674,944* | *$17,474,000* | *$19,070,000* | *$18,675,000* |

Sources: H.Rept. 118-122, pp. 96-100; S.Rept. 118-43, pp. 106-109, P.L. 118-42.

Note: Family Housing budget numbers include Family Housing Construction, the Family Housing Operation & Maintenance, Family Housing Improvement Fund (FHIF), and the Unaccompanied Housing Improvement Fund (UHIF).

## Selected Funding Items

In Section 124, the appropriations act provided $514.1 million for additional construction funds for projects listed on DOD's FY2024 unfunded priorities lists.

The appropriations included $662.4 million for planning and design and construction of new barracks facilities, also known as unaccompanied housing. That includes Section 132, in which Congress appropriated $15 million for each of the three military departments for planning and design for new barracks facilities. The joint explanatory statement that accompanied the enacted law directs each of the services to provide the defense committees with a report for how the services intend to address poor condition of barracks facilities that the Government Accountability Office (GAO) identified in a report released in September 2023 (GAO–23–105797).

The appropriations included $336.3 million to support Child Development Centers (CDCs), which provided $276.8 million for construction of six new CDCs and $59.5 million in planning and design funding to help develop future construction.

The appropriations act included $69 million for planning and design and $62 million in unspecified minor construction for projects as identified on INDOPACOM's unfunded priorities list.

**Table 2.** Selected military construction projects funded by P.L. 118-42

(in thousands of dollars of budget authority)

| *Location* | *Project Title* | *Requested Amount* | *P.L. 118-42* |
|---|---|---|---|
| Joint Base Pearl Harbor-Hickam, HI | Dry Dock 3 Replacement (Inc) | $1,318,711 | $1,318,711 |
| Portsmouth Naval Shipyard, ME | Multi-Mission Drydock #1 Extension (Inc) | $544,808 | $544,808 |
| Fort Meade, MD | NSAW Recap Building 4 (Inc) | $315,000 | $315,000 |
| Joint Region Marianas, GU | PDI: Joint Communication Upgrade (Inc) | $292,830 | $146,415 |
| Naval Weapons Station Yorktown, VA | Weapons Magazines | $221,920 | $110, 960 |
| Naval Submarine Base New London, CT | Weapons Magazine & Ordnance Operations Facility | $219,200 | $109,600 |
| March Air Reserve Base, CA | KC-46 Two Bay Maintenance/Fuel Hangar | $201,000 | $201,000 |
| Naval Base Kitsap, WA | Shipyard Electrical Backbone | $195,000 | $195,000 |
| Fort Meade, MD | Cybersecurity Operations Facility | $186,480 | $186,480 |

**Table 2.** (Continued)

| *Location* | *Project Title* | *Requested Amount* | *P.L. 118-42* |
|---|---|---|---|
| Fort Belvoir, VA | DIA Headquarters Annex | $185,000 | $185,000 |
| Guam/Joint Region Marianas, GU | Missile Integration Test Facility | $174,540 | $87,270 |
| Ellsworth Air Force Base, SD | B-21 Phase Hanger | $160,000 | $160,000 |
| Marine Barracks Washington, DC | Bachelor Enlisted Quarters | $131,800 | $65,900 |
| Naval Station Norfolk, VA | MQ-25 Aircraft Laydown Facilities | $114,495 | $57,248 |
| Port Hueneme, CA | Laboratory Compound Facilities Improvement | $110,000 | $110,000 |
| Port Hueneme, CA | Laboratory Compound Facilities Improvement | $110,000 | $110,000 |
| Marine Corps Base, HI | Water Reclamation Facility Upgrade | $0 | $109,000 |
| Fort Jackson, SC | Reception Barracks Complex, Phase 2, Cost to Complete | $0 | $101,000 |
| Joint Base Pearl Harbor-Hickam, HI | Dry Dock Replacement | $0 | $90,000 |
| Lexington Armory, NY | National Guard Readiness Center addition/alteration | $0 | $90,000 |
| Joint Base Pearl Harbor-Hickam, HI | Waterfront Production Facility (planning and design) | $0 | $49,080 |
| Fort Novosel, AL | Adv. Individual Training Barracks Complex (Cost to Complete) | $0 | $41,200 |
| Fort Hunter Liggett, CA | Network Enterprise Center | $0 | $40,000 |

Source: FY2024 Enacted from P.L. 118-47 and Senate, Congressional Record, Daily Edition, Vol. 170, no. 39, part II (March 5, 2024), p. S1246-S1263.

Notes: Projects listed in this table were selected based on a CRS longitudinal analysis of the president's budget request for FY2024, and the House- and Senate-passed drafts of the FY2024 MILCON-VA bills and P.L. 118-47. In a previous CRS Insight about the House- and Senate-passed drafts of the FY2024 MILCON-VA bills, CRS compiled three lists of projects: (1) the 10 largest MILCON projects (by dollar value) that were included in the PBR; (2) the 10 MILCON projects in the House-passed bill that showed the largest proposed change in funding (by dollar value) compared to the president's budget request; and (3) the 10 MILCON projects in the Senate-passed bill that showed the largest proposed change in funding (by dollar value) compared to the president's budget request. The table above includes all of the projects that CRS identified in that analysis of the president's budget request, and the House- and Senate-passed drafts of the FY2024 MILCON-VA bills. Acronyms include NSAW, National Security Agency Washington; PDI, Pacific Deterrence Initiative; and DIA, Defense Intelligence Agency.

## Chapter 7

# Military Construction: Better Information Sharing Would Improve DOD's Oversight*

**United States Government Accountability Office**

## Abbreviations

| | |
|---|---|
| A&E | architect-engineering |
| Corps | U.S. Army Corps of Engineers |
| DOD | Department of Defense |
| FC | Facilities Criteria |
| FEC | Facilities Engineering Command |
| HVAC | heating, ventilation, and air conditioning |
| NAVFAC | Naval Facilities Engineering Systems Command |
| NDAA | National Defense Authorization Act |
| OSD | Office of the Secretary of Defense |
| UFC | Unified Facilities Criteria |
| UFGS | Unified Facilities Guide Specifications |

## Why GAO Did This Study

For fiscal year 2025, DOD has requested over $15 billion for its military construction program, including projects ranging from child care centers to

* This is an edited, reformatted and augmented version of the United States Government Accountability Office Report to Congressional Committees, Publication No. GAO-24-106499, dated September 2024.

In: Military Construction and Appropriations
Editor: George K. Santiago
ISBN: 979-8-89530-799-1

barracks and maintenance hangars. Proper planning of such projects is critical to avoid delays from errors that could increase costs. DOD relies on UFC in completing military construction projects. At the end of fiscal year 2023, DOD had 598 military construction projects under way.

The joint explanatory statement accompanying the NDAA for fiscal Year 2022 includes a provision for GAO to review the UFC program and implementation of standards. This report examines (1) how DOD manages the UFC program, including its incorporation of relevant fiscal year 2018–2022 NDAA provisions; and (2) the extent to which DOD monitors the execution of its military construction program and projects.

GAO selected a nongeneralizable sample of five working groups and eight military construction locations. GAO analyzed relevant laws, military construction reports, and policies and procedures; and interviewed officials.

## What GAO Recommends

GAO is making seven recommendations, including for DOD to issue guidance for reporting relevant project information, and for the Army and the Navy to develop guidance and training and improve processes for sharing lessons learned. DOD concurred with all of GAO's recommendations and stated that it is taking action to implement them.

## What GAO Found

The Office of the Secretary of Defense (OSD) and the military departments (Army, Navy, and Air Force) collectively manage the Unified Facilities Criteria (UFC) program. UFC documents provide criteria for the planning, design, and construction, among other things, of Department of Defense (DOD) owned facilities. According to officials, DOD creates or updates UFC documents based on National Defense Authorization Acts (NDAA), among other things. GAO found that DOD largely incorporated relevant fiscal year 2018–2022 NDAA provisions.

GAO found that DOD does not fully monitor the execution of its military construction program and projects. OSD, which is responsible for general program oversight, collects detailed information on two military construction portfolios including the Indo-Pacific Command area of responsibility and two other high-profile projects. However, information OSD collects for all other construction projects is limited and OSD relies on Army and Navy construction agents for project monitoring. OSD officials told GAO the information they collect for these projects is for reporting purposes only and is not relevant for identifying trends, which can help inform a risk-based oversight approach. DOD's annual reports on military construction delays show that over the prior 5 fiscal years, poor initial planning contributed to about 25 percent of the projects delayed for at least a year. By issuing guidance to require reporting of more relevant information, such as the DOD construction agent responsible for each project and planning and design milestones, OSD would have better visibility into projects and could better identify and address individual and systemic issues.

Further, GAO found that Army and Navy construction agents do not consistently document and share lessons learned in their project monitoring. For example, one building was delayed for over 3 years due to design errors (e.g., incorrect roof design) and insufficient quality control oversight, according to Army documentation (see figure). Army construction agent officials said they were not using a lessons-learned system to share project observations, which may help prevent repeating past mistakes on future projects.

Example of a Delayed Military Construction Project

**Special Operations Forces (SOF) Training Command Building**

**Location:** Fort Liberty, NC

**Construction start date:** Mar. 2017

**Original planned construction completion date:** Feb. 2019

**Actual construction completion date:** July 2022

**Original construction amount:** Approximately $38.7 million

**Current obligation amount:** Approximately $54.3 million

**Scope of project:** A three-story facility that includes administrative and operations space, classrooms, and a library.

**Reasons for schedule delays:** Design errors (e.g., not including all roof loading), insufficient quality control, personnel turnover, and changing customer requirements.

Source: GAO analysis of Department of Defense information; U.S. Army Corps of Engineers (photo). | GAO-24-106499

Note: For more details, see figure 6 in GAO-24-106499.

Better guidance, training, and processes for sharing lessons learned could help prevent future mistakes, such as insufficient quality control, and save resources.

September 16, 2024

The Honorable Jack Reed
Chairman

The Honorable Roger Wicker
Ranking Member
Committee on Armed Services
United States Senate

The Honorable Mike Rogers
Chairman

The Honorable Adam Smith
Ranking Member
Committee on Armed Services
House of Representatives

For fiscal year 2025, the Department of Defense (DOD) requested over $15 billion for its military construction program, including projects ranging from child development centers and test facilities to barracks and maintenance hangars.[1] The U.S. Army Corps of Engineers (Corps) and Naval Facilities Engineering Systems Command (NAVFAC) function as construction agents for DOD facilities.[2] As such, they are responsible for the design and execution of military construction projects. According to DOD documentation, proper planning of projects is critical to avoid delays that may result from design errors and could lead to increased costs for project sponsors.

DOD's Unified Facilities Criteria (UFC) are technical criteria documents and specifications used for the planning, design, and construction of DOD military construction projects. The objectives of the UFC program include streamlining military facilities criteria, increasing reliance on private sector standards, and creating a more efficient criteria development process.

---

[1] According to DOD, the military construction program includes all work necessary to produce complete and usable facilities, or to complete usable improvements to existing facilities, in support of DOD components.

[2] Department of Defense (DOD) Directive 4270.05, Military Construction (Feb. 12, 2005) (incorporating change 1, effective Aug. 31, 2018).

UFC documents consist of around 150 issuances and provide common requirements across DOD for a wide range of standards, including for safety, sustainability, durability, and functionality of DOD facilities.

The joint explanatory statement accompanying the National Defense Authorization Act (NDAA) for Fiscal Year 2022 includes a provision for us to review the UFC program and the implementation of standards.[3] This report examines (1) how DOD manages its UFC program, including the extent to which it incorporated relevant provisions from the NDAAs for fiscal years 2018 through 2022 into UFC documents; and (2) the extent to which DOD monitors the execution of its military construction program and projects.

To address our first objective, we interviewed officials from selected working groups that develop and update UFC documents to gain their perspectives on how they maintain UFC documents within their respective areas of responsibility. We reviewed and identified working groups responsible for core UFC documents, which provide requirements for the majority of the traditional building systems prevalent on DOD facility construction projects. We then selected a nongeneralizable sample of five of these 23 identified working groups for further review based on the number of criteria change requests in fiscal year 2022 and the average number of publications within the respective areas of responsibility for each working group.[4]

To assess the extent to which DOD incorporated relevant provisions from the NDAAs for fiscal year 2018 through fiscal year 2022, we analyzed the NDAAs from these fiscal years and identified 16 total provisions related to UFC. We then requested and obtained DOD documentation on how the department implemented these requirements and analyzed the results. If DOD addressed all relevant sections of the identified provisions, we determined that it incorporated the provision. If DOD addressed at least part of the provision, we determined that DOD partially incorporated the provision. Finally, if DOD addressed none of the provision, we determined that DOD did not incorporate the provision.

---

[3] House Armed Services Comm. Print No. 2, at 1,249 (2021).

[4] A criteria change request, commonly referred to as a CCR, is a requested modification to a UFC, Unified Facilities Guide Specification, or Facilities Criteria. Criteria change requests may be provided by anyone, such as but not limited to industry representatives, private entities, or DOD personnel. The requests are automatically routed for adjudication to the appropriate criteria managers within working groups. DOD Military Standard 3007G, *Department of Defense Standard Practice: Unified Facilities Criteria, Facilities Criteria and Unified Facilities Guide Specifications* (Nov. 1, 2019).

To address our second objective, we reviewed monthly and annual military construction reports to determine what information was included and reasons DOD identified and noted for delayed military construction projects. We found the data in these reports to be sufficiently reliable for the purposes of selecting a subset of military construction projects for further review. We reviewed DOD policies, procedures, and documentation, and interviewed officials to determine how they conduct oversight. We then selected a nongeneralizable sample of eight military construction projects based on specific criteria.

Specifically, we selected four military construction projects within the United States that DOD identified as being delayed for at least 365 days in DOD's fiscal year 2022 report on delayed military construction projects. We analyzed this report and selected military construction projects that were delayed, at least in part, due to a design-related factor. From this subset we selected two military construction projects from the Corps and two projects from NAVFAC based on the length of schedule delays and the description of the design issue. We also selected four military construction projects the Corps and NAVFAC identified as being successful.[5] Specifically, we selected one project that cost over $50 million and one project that cost under $50 million for both the Corps and NAVFAC. We analyzed these projects by assessing design documentation and interviewing responsible officials. See appendix I for a complete list of organizations and military construction projects we selected and reviewed.

We conducted this performance audit from January 2023 to September 2024 in accordance with generally accepted government auditing standards. Those standards require that we plan and perform the audit to obtain sufficient, appropriate evidence to provide a reasonable basis for our findings and conclusions based on our audit objectives. We believe that the evidence obtained provides a reasonable basis for our findings and conclusions based on our audit objectives.

---

[5] For the purposes of this report, we excluded military construction projects where the Air Force Civil Engineer Center served as the DOD construction agent because the Corps and NAVFAC serve as the DOD construction agent for military construction worldwide, except for the British Isles.

# Background

## Legislative Requirement for Facility Design Criteria

According to Military Standard 3007G, House Report 105-247 accompanying the Military Construction Appropriations Act, 1998 directed DOD and the military services to establish procedures for unification of facilities criteria.[6] The three military departments, under the auspices of the Office of the Secretary of Defense (OSD), established a working group to address unification issues. The working group surveyed procedures, criteria, and guide specifications; evaluated criteria uniformity; identified areas where greater uniformity was practical; analyzed management options for a uniform guidance system; and submitted a framework to the congressional defense committees. DOD developed and published Military Standard 3007G to implement that framework.[7]

# DOD's Criteria Framework

According to guidance, DOD's criteria framework was designed to standardize and streamline the process for developing, maintaining, and disseminating criteria across the department.[8] The framework provides facility planning, design, construction, operation and maintenance, sustainment, restoration, and modernization criteria for DOD-owned facilities. In addition, DOD relies on industry standards to facilitate the implementation, management, and execution of criteria to the maximum extent practicable.[9] DOD's criteria framework consists of the following:

---

[6] DOD Military Standard 3007G; *see* H.R. Rep. No. 105-247, at 8 (1997).

[7] See DOD Military Standard 3007G. According to Corps officials, the first iteration of Military Standard 3007G, now superseded, was published as Military Standard 3007 in 2000.

[8] DOD Military Standard 3007G.

[9] See DOD Military Standard 3007G. According to the standard, the unification process for the criteria maximizes the use of nongovernment standards in accordance with law and guidance. Specifically, the National Technology Transfer and Advancement Act of 1995 requires all federal agencies and departments, with some exceptions, to use technical standards that are developed or adopted by voluntary consensus standards bodies, using such technical standards as a means to carry out policy objectives or activities determined by the agencies and departments. Pub. L. No. 104-113, § 12(d) (1996) (as amended). Additionally, Office of Management and Budget Circular A-119, *Federal Participation in the Development and Use of Voluntary Consensus Standards and in Conformity Assessment*

- *Unified Facilities Criteria (UFC)* provide DOD facility design requirements. "Core" UFC documents consist of those UFC that address general building requirements, high-performance building mandates, and discipline-specific systems found in most buildings.[10] The core UFC address building code compliance, life safety, legislation compliance, and performance.[11] The core UFC provide requirements for the majority of traditional building systems prevalent on DOD facility construction projects. UFC 1-200-01, DOD Building Code, is the foundational document of the UFC program. It provides general building requirements and overarching criteria; establishes the use of consensus building codes and standards; establishes criteria implementation rules and protocols (including core UFC); and identifies unique military criteria.[12] It serves as the bridge between industry standards and DOD criteria.
- *Facilities Criteria (FC)* define functional requirements for specific types of facilities within a military service. Such facilities include child development centers, barracks, or dining facilities. Requirements are established by the service regulatory or functional authority rather than the technical community of architects and engineers. FC are nonunified criteria and address a service's unique requirements.
- *Unified Facilities Guide Specifications (UFGS)* provide common facility specifications, standards and engineering practices for the DoD and other supported agencies. UFGS are, according to NAVFAC officials, for the purpose of translating design criteria into construction contract requirements. The officials stated that UFGS are edited by the designer of record for use on each specific project as the basis of construction contract documents. UFGS that are

---

*Activities* (Jan. 27, 2016), directs agencies to use voluntary consensus standards in lieu of government-unique standards except where inconsistent with law or otherwise impractical.

[10] The term "high-performance building" means a building that integrates and optimizes on a life-cycle basis all major high-performance attributes, including energy conservation, environment, safety, security, durability, accessibility, cost-benefit, productivity, sustainability, functionality, and operational considerations. The term "high-performance green building" means a high-performance building that, during its life cycle, as compared with similar buildings, reduces energy, water, and material resource use and reduces negative impacts on the environment throughout the life cycle of the building, among other things. See 42 U.S.C. § 17061.

[11] DOD Military Standard 3007G.

[12] DOD, Unified Facilities Criteria (UFC) 1-200-01, *DOD Building Code* (Sept. 1, 2022) (incorporating change 3, effective Feb. 26, 2024).

frequently used but specialized (tied to a mission, such as runways, or specialized components or systems) are considered 'specialty' UFGS. See figure 1 below.

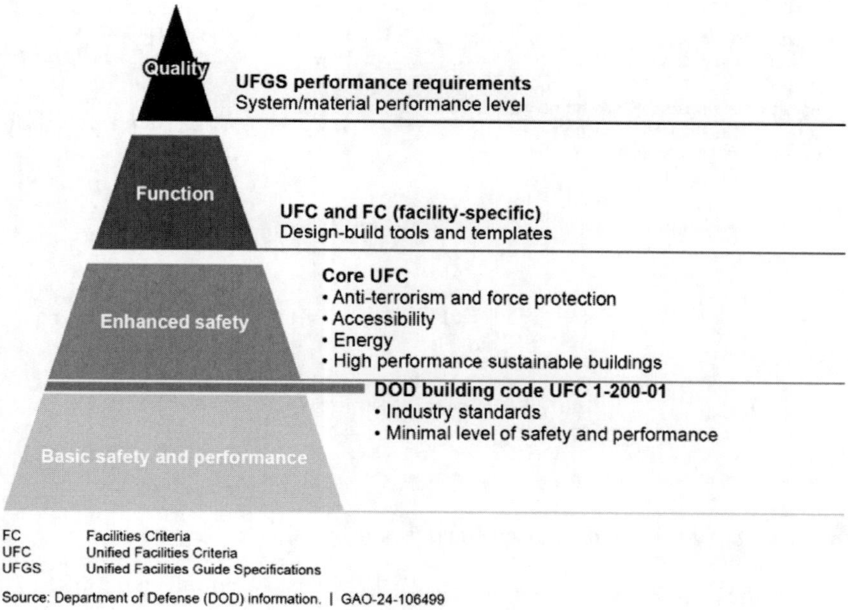


**Figure 1**. Overview of DOD's Criteria Framework.

## Military Construction Process and Key Project Stakeholders

The military construction process ranges from facility planning (conducted at each installation that has a requirement for new facilities); through project programming and budgeting (in DOD component commands and headquarters); to project design and construction (via the department's designated design and construction agents—the Corps and NAVFAC).

OSD exercises general oversight over the military construction program. Specifically, the Office of the Assistant Secretary of Defense for Energy, Installations, and Environment is responsible for monitoring the program to ensure that construction agents execute projects in the most efficient, expeditious, and cost-effective manner possible.[13]

[13] See DOD Directive 4270.05. The directive assigns these responsibilities to the Deputy Under Secretary of Defense for Installations and Environment. According to DOD officials, this

Each military construction project has three primary types of stakeholders, as shown in figure 2.[14]

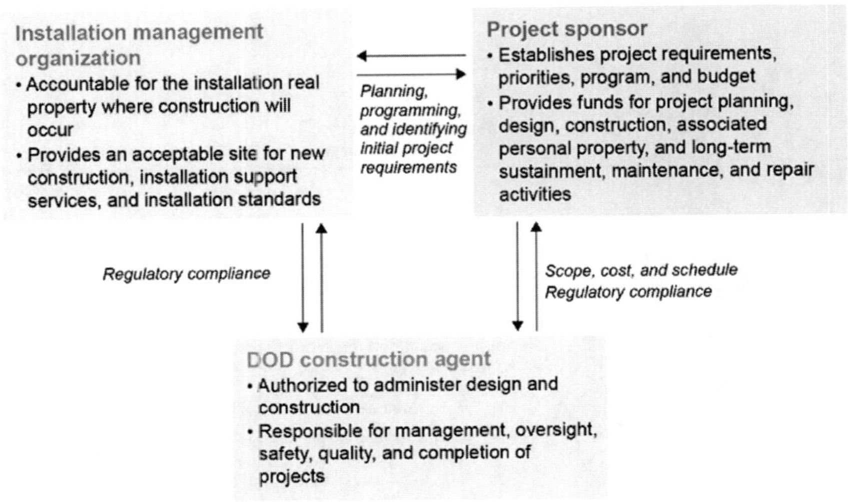


Source: Department of Defense (DOD) information. | GAO-24-106499

**Figure 2.** Overview of Military Construction Key Stakeholder Responsibilities.

- *Installation management organization.* The military department (or Washington Headquarters Services with respect to the Pentagon Reservation) with custody and accountability for the physical infrastructure and real property of the military installation on which construction will occur. It has a variety of responsibilities including provision and characterization of an acceptable site for new construction, provision of installation support services, and governance of installation standards—such as installation-specific

position no longer exists and military construction program oversight duties are currently performed by the Assistant Secretary of Defense for Energy, Installations, and Environment.

[14] Principal Deputy Assistant Secretary of Defense for Energy, Installations, and Environment Memorandum, *Military Construction Project Key Stakeholders and Responsibilities* (Mar. 25, 2022). For any given project, each project stakeholder may be a different DOD component, but nothing prohibits a single DOD component serving as one or more project stakeholders for the same project. DOD components include the following: OSD; the Chairman of the Joint Chiefs of Staff and the Joint Staff; the military departments, the defense agencies; DOD field activities; the Combatant Commands; and Washington Headquarters Services.

architectural and design requirements. It also coordinates with the project sponsor and tenant organization.

- *Project sponsor.* The DOD component that establishes the project requirements, priorities, program, and budget. The project sponsor's responsibilities include developing the project (including planning and preliminary design); providing funds for project planning, design, construction, associated personal property (including collateral equipment), and long-term sustainment, maintenance, and repair activities; and establishing, advocating for, and defending (before OSD and Congress) the project requirements, scope, budget, schedule, justification, and funding source. Further, the project sponsor is responsible for fiscal compliance within the scope of its duties, including congressional notifications and reprogramming actions.
- *DOD construction agent.* The DOD organization authorized, according to DOD guidance, pursuant to section 2851 of title 10, U.S. Code, and DOD policy, to administer design and construction contracts for DOD facilities projects in accordance with applicable laws and regulations.[15] The DOD construction agent is responsible for administration, management, oversight, safety, quality, and completion of in-house and contracted facilities design and construction activities.[16] It is also responsible for ensuring facilities design and construction are consistent with authorized project scope and conform with DOD UFC and project sponsor requirements. Typically, the Corps is the construction agent for Army-funded military construction projects and NAVFAC is the construction agent for Navy and Marine Corps-funded military construction projects. The Department of the Air Force may either use the Corps or NAVFAC as the construction agent for its projects.
    - Corps. The Corps' military missions area includes military construction and is organized geographically into eight divisions that are assigned to the military construction mission.[17] These are

---

[15] Principal Deputy Assistant Secretary of Defense for Energy, Installations, and Environment Memorandum, *Military Construction Project Key Stakeholders and Responsibilities* (Mar. 25, 2022).

[16] In-house design refers to when the Corps or NAVFAC officials create construction project design documents.

[17] These are the Northwestern Division, South Pacific Division, Southwestern Division, Great Lakes and Ohio River Division, North Atlantic Division, South Atlantic Division, Pacific Ocean Division, and the Transatlantic Division.

broken down into 25 military programs districts. Districts produce various products including design documents, studies, and programming estimates.

- NAVFAC. NAVFAC comprises two regional commands—NAVFAC Atlantic in Norfolk, Virginia, and NAVFAC Pacific in Pearl Harbor, Hawaii. NAVFAC Atlantic aligns operationally with the Navy's Atlantic Fleet and oversees four Facilities Engineering Commands (FEC).[18] FECs are responsible for the integration of planning, programming, and delivery of NAVFAC capabilities and functions to supporting commands. NAVFAC Pacific aligns operationally with the Navy's Pacific Fleet and oversees five FECs.[19]
- Air Force Civil Engineer Center. The Air Force Civil Engineer Center is the designated DOD construction agent for military construction projects in the British Isles. Also, in some cases, the Air Force Civil Engineer Center may design and construct Air Force projects where the Air Force and the commander of the assigned construction agent agree it is the most efficient, expeditious, and cost-effective means to complete the project.

## Design Process

Under DOD guidance and criteria, design may occur through in-house professional staff or through a contract with an architect-engineering (A&E) firm. The designer of record is responsible for architectural and engineering aspects of the project and for the design work for each design discipline (e.g., electrical, mechanical). The Corps or NAVFAC is the DOD construction agent for military construction projects, unless, according to DOD officials, otherwise designated by the Assistant Secretary of Defense for Energy, Installations, and Environment. In cases of in-house design, the Corps or NAVFAC is also the designer of record. In cases when DOD contracts with an A&E firm, the firm is the designer of record.

The design process includes the development of design plans, design specifications, design submittals, cost estimates, and extensive site

---

[18] The FECs that are aligned with NAVFAC Atlantic are NAVFAC Europe Africa Central, NAVFAC Mid-Atlantic, NAVFAC Southeast, and NAVFAC Washington.

[19] The FECs that are aligned with NAVFAC Pacific are NAVFAC Far East, NAVFAC Hawaii, NAVFAC Marianas, NAVFAC Northwest, and NAVFAC Southwest.

investigation.[20] The preliminary design phase is normally the critical point in design and is called the '35-percent' stage because the project design should provide sufficient detail to define the scope, criteria, and cost estimates for budget consideration by the time it is 35 percent complete. Additionally, under Corps guidance, design documents are reviewed typically at the 35-percent completion, 65-percent completion, and 95- percent completion final design stages.[21] NAVFAC design submissions, including quality control reviews, are to occur at points when the design is typically about 35-percent to 50-percent complete, 100-percent pre-final phase, and final design.[22] According to DOD guidance, DOD initiates the final design phase after thorough review of design submissions to ensure that preliminary design documents properly addressed requirements and criteria.

NAVFAC and Army Corps officials said that if an A&E firm is designing the project, then the firm provides quality control on their design deliverables at each submittal phase and NAVFAC or Corps designers provide quality assurance over those deliverables.[23] In contrast, according to the officials, if the project is an in-house NAVFAC or Corps design, then NAVFAC or the Corps provides quality assurance and quality control over their own design submittals, with separate teams for each type of review.

The design phase typically can take from 1 to 2.5 years. According to the DOD Financial Management Regulation, upon completion of the work drawings, contract specifications, and bidding documents, the project is ready to be advertised for construction.[24] According to the regulation, major construction projects require both congressional authorization and appropriation before a construction contract can be awarded.[25] Figure 3 summarizes this process.

---

[20] See DOD 7000.14-R, Financial Management Regulation, vol. 3, ch. 17, Accounting Requirements for Military Construction Projects, Appx. 1 (July 2021).

[21] Army Engineer Regulation 1110-3-12, *Military Engineering and Design: Quality Management* (Mar. 25, 2021).

[22] DOD Facilities Criteria (FC) 1-300-09N, *Navy and Marine Corps Design Procedures* (May 17, 2024). The 100-percent design submittal is the pre-final stage during which the intent is to provide a complete set of design deliverables.

[23] The Corps states that quality control includes processes to ensure that project performance meets agreed-upon stakeholder requirements that are consistent with law, regulations, policies, sound technical criteria, schedule, and budget. Quality assurance includes those processes to ensure that quality control activities are being accomplished and are effective in producing a product that meets required quality metrics. Army Engineer Regulation 1110-3-12.

[24] DOD 7000.14-R, vol. 3 ch. 17.

[25] The regulation states that major military construction projects include all acquisition or construction, additions, expansions, extensions, conversions, alterations, or replacements of

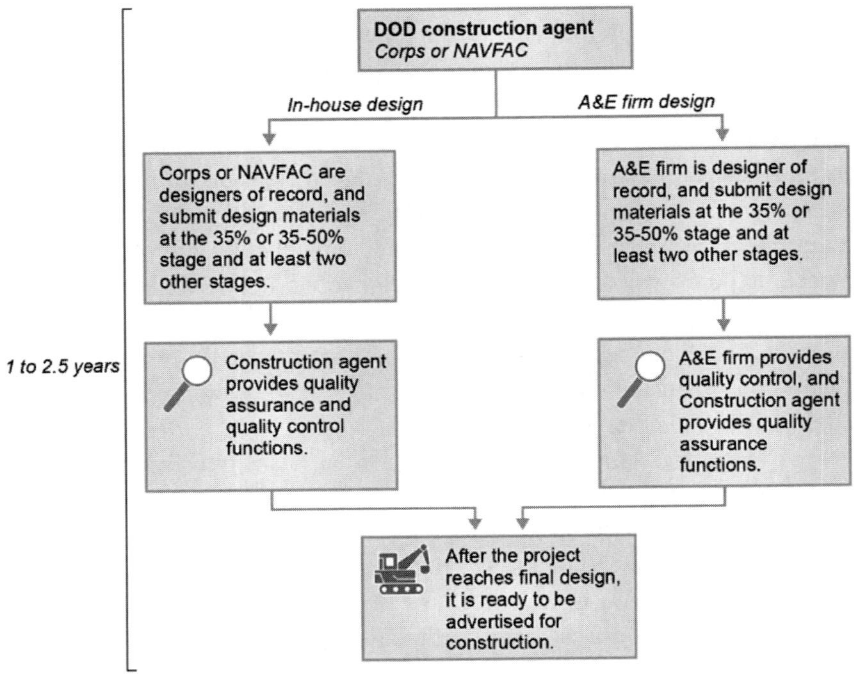


A&E Architect and engineering
Corps U.S. Army Corps of Engineers
NAVFAC Naval Facilities Engineering Systems Command

Source: Department of Defense (DOD) information; GAO (icons). | GAO-24-106499

**Figure 3.** Overview of Typical Military Construction Design Process.

## Statutory Reporting Requirements

Section 2851 of title 10, U.S. Code, requires the Secretary of Defense to maintain an internet site with information on military construction projects, such as those military construction projects or military family housing projects that have been specifically authorized by an act of Congress.[26] Specifically,

facilities with costs in excess of a certain amount ($9 million as of July 2024) or any project, regardless of cost, approved as a specific line item in the military construction budget request. See DOD 7000.14-R, vol. 3, ch. 17.

[26] 10 U.S.C. § 2851. Other projects for which the Secretary of Defense is required to maintain information on the internet site are (1) each project carried out with funds authorized for the operation and maintenance of military family housing; (2) each project carried out with funds authorized for the improvement of military family housing units; (3) each unspecified minor construction project carried out under the authority of section 2805(a) of title 10, U.S. Code; (4) each military department project with a total cost in excess of $15 million for Facilities Sustainment, Restoration, and Modernization; and (5) each military

the provision requires the Secretary of Defense to provide the following information for each project:

- The solicitation date and award date (or anticipated dates) for each contract entered into (or to be entered into) by the United States in connection with the project.
- The contract recipient, contract award amount, construction milestone schedule proposed by the contractor, and construction completion date stipulated in the awarded contract.
- The most current DOD Form 1391, Military Construction Project Data, for the project.
- The progress of the project, including the percentage of construction currently completed and the current estimated construction completion date.
- The current contract obligation of funds for the project, including any changes to the original contract award amount.
- If funds appropriated for the project have been diverted for use in another project, the project to which the funds were diverted and the amount so diverted.
- For accounts such as planning and design, unspecified minor construction, and family housing operation and maintenance, detailed information regarding expenditures and anticipated expenditures under these accounts and the purposes for which the expenditures are made.

The provision requires the Secretary of Defense to update the information as promptly as practicable, but not less frequently than once a month, to ensure that the information is available in a timely manner. Additionally, the provision requires the Secretary of Defense to submit to the Committees on Armed Services of the House of Representatives and Senate annual reports on each military construction project or military family housing project for which, as of the end of the most recent fiscal year, the estimated completion date of the project is more than 1 year later than the date proposed at the time DOD awarded the contract.

---

construction project, military department Facilities Sustainment, Restoration, and Modernization project, or military family housing project regarding which a statutory requirement exists to notify Congress.

## Multiple DOD Entities Manage the UFC Program and DOD Has Largely Incorporated NDAA Provisions

### OSD and the Military Departments Collectively Manage DOD's UFC Program

OSD and the military departments work together to manage the UFC program. Within the program, working groups comprising representatives from offices within OSD and the Departments of the Army, Navy, and Air Force develop and maintain UFC documents. Working groups can either be "discipline" (meaning they align with the disciplines of architecture and engineering, such as electrical and mechanical disciplines) or "functional" (meaning they align with functional or operational areas such as sustainability and aviation). We refer to both types as "working groups" in this report and list all identified working groups in appendix II.

The Engineering Senior Executive Panel is responsible for directing and funding the UFC program, while the Coordinating Panel oversees the day-to-day operations of the UFC program and the working groups.[27] Working groups are responsible for a portfolio of UFC documents within their respective areas of expertise, as illustrated in Figure 4.[28]

DOD can revise or change UFC documents. For example, according to DOD officials, UFC-1-200-01, DOD Building Code, is required to be revised every 3 years. Revisions require Engineering Senior Executive Panel review and approval. Working groups can make changes to UFC documents, requiring less internal review than revisions.[29] Revisions and changes are required to be incorporated into military construction projects up until 35 percent of design completion, and in other certain cases, such as in some circumstances when there is a delay of more than 18 months between design completion and the solicitation of offers for construction.[30] Working group officials also stated

---

[27] The Engineering Senior Executive Panel is a four-member panel composed of a Senior Executive Service representative from each military department and OSD. The Coordinating Panel is a separate four-member panel composed of a senior representative of each military department and OSD.

[28] The Engineering Senior Executive Panel approves UFC issuances and reissuances. Officials told us that the Engineering Senior Executive Panel has delegated responsibility for approving UFGS changes to the working groups.

[29] The Coordinating Panel approves changes to UFC documents.

[30] DOD Military Standard 3007G. Specifically, the standard requires projects with a delay of more than 18 months to be re-evaluated to determine if any design revision is necessary due to changes in criteria or site infrastructure.

that UFC document changes that affect life and safety may still be incorporated later in the project. Overall, from the time DOD changes or updates a UFC it could take several years before a completed construction project reflects the new criteria.

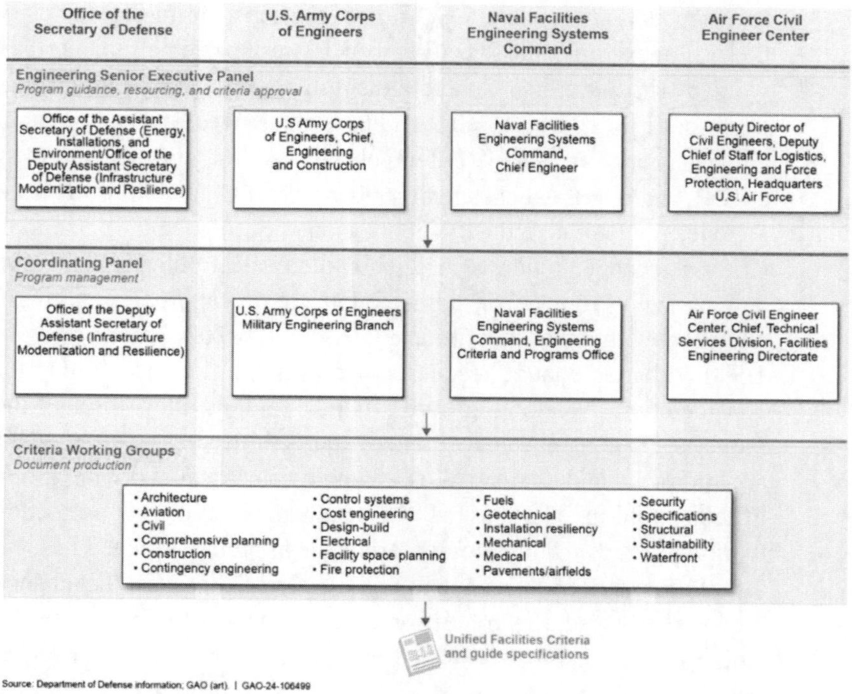


Source: Department of Defense information; GAO (art). | GAO-24-106499

**Figure 4.** Overview of Unified Facilities Criteria Program Management.

According to working group officials, the three main inputs for DOD to create new criteria or update existing criteria are as follows:

- Laws, regulations, and policy. Officials told us the Coordinating Panel tasks working groups with providing comments on draft legislative language and implementing enacted NDAA provisions by making changes to criteria documents. Officials told us that comments on previous draft legislative language have been incorporated into enacted NDAA provisions. Additionally, officials told us that executive orders and federal government regulations, such as those issued by the Department of Energy, are incorporated into criteria.

- Change in industry standards. Industry standards are nongovernment standards prepared by nationally and internationally recognized technical, professional, and industry associations and societies. Officials told us they are aware of, and prepare for, industry standard issuance cycles.[31] Officials also told us they work with industry representatives to ensure UFC reflect the most up-to-date industry codes. For example, the structural working group officials told us they worked with American Concrete Institute representatives to develop a new code for concrete slabs to address challenges DOD was facing.
- Criteria change request. A criteria change request, also referred to as a CCR, is a requested change or revision to UFC, UFGS, or FC from anyone, including industry representatives or DOD personnel. Criteria change requests are submitted electronically and are automatically routed for adjudication to the appropriate criteria managers within working groups. In fiscal year 2023, approximately 1,500 criteria change requests were submitted to DOD for consideration. Working group officials told us they discuss the requests and consult subject matter experts, if necessary, during their adjudication deliberations. For example, electrical working group officials told us they asked the mechanical and fire protection working groups for input on criteria change requests related to UFC 3-520-05, Stationary And Mission Batteries, because the change requests involved fire protection and ventilation for battery areas. Additionally, electrical working group officials told us they provided input to the mechanical working group on criteria change requests related to the control of electronically controlled motors for heating, ventilation, and air conditioning (HVAC) systems.

## DOD Has Largely Incorporated Provisions for Fiscal Years 2018– 2022 and Plans to Incorporate Policy-Related Provisions

We found that, based on information from DOD, DOD incorporated 13 provisions and partially incorporated two provisions of the 16 relevant NDAA provisions from fiscal year 2018 through fiscal year 2022 into UFC documents, other guidance, or processes; and is planning to address one

---

[31] Officials told us industry standards are updated and reissued on a recurring basis, such as every 3 years.

provision.[32] For example, DOD incorporated NDAA provisions pertaining to issuing criteria for installing microgrids (i.e., networks of standby power systems in the event of commercial power system failure) and criteria to include the consideration for solar roofing into UFC documents.[33]

For the two partially incorporated NDAA provisions, DOD plans to take additional actions beyond those completed to date.[34] For example, DOD officials stated that although the military services have guidance on the consideration of potential long-term adverse environmental effects, DOD officials plan to create a new UFC document by February 2025 that will more comprehensively incorporate section 2805 of the NDAA for Fiscal Year 2020. Specifically, this section requires certain modifications to DOD Form 1391, Military Construction Project Data—which DOD submits to Congress in support of funding requests for military construction projects—certifying that proposed military construction projects consider certain potential long-term adverse environmental effects.

DOD has not incorporated one NDAA provision related to the determination of, and notification to Congress about, when executive orders will affect the cost or scope of work for military construction projects.[35] Officials told us that they have plans to issue policy to address the provision and are working to determine a methodology for identifying how to isolate the information within a military construction project. See appendix III for relevant excerpts from and our full analysis of the 16 relevant NDAA provisions from fiscal years 2018 through 2022.

OSD officials told us some NDAA provisions can have unintended consequences. For example, these officials stated that a provision requiring, with some exceptions, a specific type of airfield runway lighting resulted in

---

[32] For purposes of this report, relevant NDAA provisions were those enacted provisions either requiring specific changes to UFC or related documents, or other changes to DOD documents or policy that were being implemented through UFC amendments. UFC documents include UFC and UFGS documents.

[33] See James M. Inhofe National Defense Authorization Act for Fiscal Year 2023, Pub. L. No. 117-263, §§ 2810-11 (2022). Specifically, DOD officials told us that the microgrid provision was incorporated through the publication of UFC 3-550-04, *Resilient Installation Microgrid Design* (Mar. 1, 2024) and UFGS 26 37 13, *Microgrid Control System* (February 2024), while the solar roofing provision was incorporated in the February 2024 change to UFC 1-200-01.

[34] These two provisions include section 2805 of the John S. McCain National Defense Authorization Act for Fiscal Year 2019, Pub. L. No. 115-232 (2018) and section 2805 of the National Defense Authorization Act for Fiscal Year 2020, Pub. L. No. 116-92 (2019).

[35] See James M. Inhofe National Defense Authorization Act for Fiscal Year 2023, Pub. L. No. 117-263, § 2812 (2022).

the lights being regularly damaged during snow plowing.[36] They also told us some NDAA provisions requiring changes to the UFC would be better implemented through DOD policy instead of a technical change. For example, OSD officials told us an NDAA for Fiscal Year 2022 provision requiring a UFC be amended to require the inclusion of lactation rooms in military construction planning and design would be better addressed in policy because the provision is unrelated to building specifications or codes.[37]

To address congressionally mandated UFC changes that are policy oriented, OSD officials told us they started to develop a requirements- based UFC document in fiscal year 2024 that will eventually serve as a repository for all project-related requirements that do not contain technical design criteria. According to the officials, installation planners will use this document to ensure Congress' intent for specific building features can be included in the initial facility requirement documents.

## DOD Does Not Fully Monitor the Execution of Its Military Construction Program and Projects

OSD gathers some information on military construction projects, but it does not have information on all projects that would better enable it to fully monitor the program. In addition, DOD construction agents do not consistently document and share observations and lessons learned as they monitor military construction projects.

### OSD Is Limited in Its Ability to Monitor Military Construction Program Execution

In overseeing military construction program execution, OSD gathers detailed information on some portfolios and high-profile projects. OSD also gathers limited information on all other projects to meet congressional reporting requirements. However, OSD does not have sufficient relevant information on

[36] See National Defense Authorization Act for Fiscal Year 2018, Pub. L. No. 115-91, § 2872 (2017).

[37] See National Defense Authorization Act for Fiscal Year 2022, Pub. L. No. 117-81, § 2841 (2021).

all military construction projects that would better enable it to conduct meaningful oversight over the program.

OSD oversees selected military construction portfolios and high-profile projects through periodic information gathering and briefings. OSD officials stated that they collect information based on factors such as congressional interest and high-value projects. Specifically, as of July 2024 OSD collects detailed information on two military construction portfolios: the Energy Resilience Conservation Investment Program and Indo-Pacific Command based projects. Additionally, DOD collects information on two other projects: the Presidential Aircraft Replacement Hangar and the Joint Intelligence Analysis Center in the United Kingdom.[38]

For these selected areas, OSD collects military construction project information that includes the DOD construction agent responsible for the project, the authorized construction amount, and current working construction estimate. OSD also collects project milestone information, such as the completion date of the 35-percent design phase, construction start date, and challenges that may cause project delays.

However, according to DOD officials, the information OSD gathers for all other military construction projects is limited to what it is required to report and is less detailed than what the office collects for overseeing selected portfolios and high-profile projects. Specifically, DOD is required to publicly report the status of certain ongoing military construction projects monthly.[39] Officials told us they only use this information to meet DOD's reporting requirement and do not use the information to conduct oversight.

To obtain information for reporting, OSD requires project sponsors to submit military construction data monthly. Reported data include the project sponsor and location as well as the original and current estimated construction completion date and contract amounts. OSD publicly reports these data on the internet. In addition to these monthly reports, OSD compiles these data into a required annual report for Congress on military construction projects with construction schedule delays over 365 days past the original construction completion date.[40] The annual report includes additional narrative describing the reason(s) for delays.

---

[38] OSD also collects information on host-country-funded projects within the Republic of Korea and Japan. Officials told us they collect this information due to the importance of the projects to the respective combatant commands.

[39] 10 U.S.C. § 2851(c).

[40] 10 U.S.C. § 2851(e). Specifically, the statute states that the Secretary of Defense shall submit to the Committees on Armed Services of the House of Representatives and Senate a report

According to the monthly report from September 2023, there were approximately 598 ongoing military construction projects at the end of fiscal year 2023 worth over $28 billion. Further, according to DOD's fiscal year 2023 annual report on military construction delays, there were 158 military construction projects delayed at least 365 days. This report states that project delays were caused by multiple factors including residual supply chain delays and shortages caused by the COVID-19 pandemic, poor initial planning for utility connections, design errors, delayed government approvals, refinement of customer requirements, and contractor issues such as poor performance. The report also said that several projects were delayed for multiple reasons.

DOD has consistently categorized the top three contributors of project delays as (1) unforeseen conditions, (2) contractor issues, and (3) poor initial planning, as shown in table 1.[41]

DOD Directive 4270.5 indicates that the Assistant Secretary of Defense for Energy, Installations, and Environment should monitor the execution of the military construction program to ensure the most efficient, expeditious, and cost-effective accomplishment of the program by DOD construction agents.[42] In addition, Standards for Internal Control in the Federal Government states that management should use quality information to achieve an entity's objectives. Such information is appropriate, current, complete, accurate, accessible, and provided on a timely basis. To do this, the standards state that management obtains relevant data from reliable internal and external sources in a timely manner and processes the obtained data into quality information. It uses this information to make informed decisions and evaluate the entity's performance in achieving key objectives and addressing risks, as OSD would do as part of its program oversight.

---

by March 1 annually on each military construction project or military family housing project for which, as of the end of the most recent fiscal year, the estimated completion date is more than 1 year later than the completion date proposed at the time the contract for the project was awarded.

[41] According to DOD's fiscal year 2022 annual report, unforeseen delays include environmental mitigation and supply chain or material shortages. Poor initial planning includes factors that could have been reasonably addressed before a delay occurred, such as the need for utility connections, design errors, insufficient planning for government approvals, and further specification of customer requirements. Contractor issues include poor performance.

[42] See DOD Directive 4270.5.

**Table 1.** Top reported contributors to military construction project delays, fiscal years 2019–2023

| Top three contributors to military construction project delays | | | | |
|---|---|---|---|---|
| Fiscal year | Total projects delayed 365 days or more | Unforeseen conditions | Contractor issues or delays | Poor initial planning |
| 2023 | 158 | 42% | 32% | 26% |
| 2022 | 109 | 53% | 22% | 25% |
| 2021 | 116 | 44% | 28% | 28% |
| 2020 | 97 | 47% | 29% | 24% |
| 2019 | 73 | 17% | 24% | 22% |

Source: Department of Defense annual reports on delayed military construction projects from fiscal years 2019–2023. | GAO-24-106499.

Note: Top three contributors do not equal 100 percent in fiscal year 2019. In DOD's fiscal year 2019 annual report on military construction delays DOD reported that most projects appeared to have unique circumstances associated with each delay with no attributable overarching category or trends associated with the delay. According to DOD's fiscal year 2022 report, unforeseen delays include environmental mitigation and supply chain or material shortages. Poor initial planning includes factors that could have been reasonably addressed before a delay occurred, such as the need for utility connections, design errors, insufficient planning for government approvals, and further specification of customer requirements. Contractor issues include poor performance. The projects listed by fiscal year are not mutually exclusive. The same project could be listed in in multiple years.

However, we found that OSD is limited in its ability to oversee the entire military construction program because it does not have sufficient relevant information on all projects that would enable it to make informed decisions and adequately address risks. As discussed above, OSD collects detailed information on some military construction portfolios and projects, which it selects based on factors such as congressional interest, high-value projects, and projects with significant or extraordinary challenges. OSD officials stated that they rely on the Corps and NAVFAC to monitor the performance of all other military construction projects. OSD also collects limited information on all other military construction projects to meet congressional reporting requirements. However, officials told us that they do not use the monthly reports and annual reports for oversight purposes and the reports do not have the relevant data they would need to identify trends within the military construction program to take corrective actions.

For example, OSD does not collect information on the DOD construction agent responsible for each military construction project, planning and design milestones, or detailed data on funds spent—information that it collects for selected military construction portfolios and projects. OSD officials stated that the types of information collected for these selected projects are helpful in conducting their program oversight responsibilities. For example, officials stated that when significant challenges arise on a project, OSD helps determine a workable way forward. Having sufficient relevant information could help OSD inform its oversight approach and ensure that it is adequately identifying and addressing risks and challenges.

As shown in the annual reports on military construction delays, over the previous 5 fiscal years, poor initial planning factors that could have been reasonably addressed before a delay occurred (including design errors) contributed to delays in approximately 25 percent of the military construction projects delayed for at least a year. Given the significant number of military construction delays and the potential for increased costs, a proactive oversight approach would benefit OSD.

OSD already collects data from project sponsors monthly and as a result, it could use existing methods to collect more detailed and relevant information to better enable monitoring of the military construction program. By issuing guidance to require reporting of more relevant information, OSD would have better visibility into projects and could better address individual project and systemic issues as they arise. OSD will also be better able to conduct program-level trend analysis to help ensure military construction projects are completed in an efficient, expeditious, and cost-effective manner. Further, OSD will be better able to identify program-level issues that contribute to the significant number of projects delayed at least a year.

## Corps and NAVFAC Inconsistently Document and Share Lessons Learned When Monitoring Projects

### Corps

We found that the Corps does not consistently document and share observations and lessons learned from its military construction projects.[43] An after-action review documents the overall performance or level of success achieved on a project at the end of each critical phase of a project to determine if objectives have been met.[44] This allows the organization or team to identify gaps and corrective actions and develop lessons learned.

---

[43] According to Army Regulation 11-33, a lesson learned is a resolved issue or best practice that improves operations or activities and results in an internalized change to capability, process, or procedure. Army Regulation 11-33, *Army Lessons Learned Program* (July 28, 2022).

[44] See Army Engineer Manual 5-1-11, Project Delivery Business Process (Sept. 1, 2022). The major phases of a military project where after-action reviews must be performed are

- Documenting observations. We observed at three out of the four Corps site visits that officials did not consistently document project observations. For example, during our site visit for the delayed Wright- Patterson Air Force Base Firehouse/Crash Rescue Station project, a Corps official told us officials did not complete an after-action review during the project or after it was completed. Corps documentation showed that this project was delayed 287 days due to design errors and associated revisions for a pre-engineered metal building. DOD's fiscal year 2022 annual report to Congress on military construction delays also cited these design delays in addition to manufacturer shipping delays. Corps officials said that other delays were due to labor shortages and material supply chain delays from COVID-19. This project began construction in May 2020 and was scheduled to complete construction in September 2021. However, the actual construction completion date was in November 2022.

  According to Corps officials, the construction contractors discovered errors and submitted a request for information to the Corps in February 2021, expressing concern about the snow and wind drift deflection calculations as referenced in the Structural Engineering UFC.[45] As a result, officials concluded that the building design was stiffer than required and the contractors expressed concern about the performance of the construction due to building movements. Corps officials said that the design did not consider the flexibility of the metal building with shifting wind conditions and that this could eventually cause the building to collapse. As a result, Corps officials said that they had to issue a contract modification to create a second design for the Firehouse/Crash station project, resulting in increased cost of around $1 million and in schedule delays (see fig. 5).

---

planning charrette, design, construction, and 9-month post completion inspection. Army Engineer Regulation 1110-3-12, *Military Engineering and Design Quality Management* (Mar. 25, 2021).

[45] See UFC 3-301-01, *Structural Engineering* (Apr. 11, 2023) (incorporating change 1, effective Oct. 2, 2023). Officials told us that when construction contractors believe they have discovered a discrepancy in design documentation, the contractors generally contact the DOD construction agent to seek clarification and a solution, if needed.

**Fire/Crash Rescue Station**

**Location:** Wright-Patterson Air Force Base, OH

**Construction start date:** May 2020

**Original planned construction completion date:** September 2021

**Actual construction completion date:** November 2022

**Original construction amount:** Approximately $6.6 million

**Current obligation amount:** Approximately $7.8 million

**Scope of project:** The project is a fire/crash rescue station that supports emergency response to 445 Airlift Wing C-17 aircraft operations. The building has two adjacent vehicle bays and additional areas for operations, training, sleeping, dining, and recreation.

**Reasons for schedule delays:** The fire/crash rescue station's schedule delays are attributed to design errors and factors related to its re-design, labor shortages, and material supply chain delays from COVID-19.

Source: GAO analysis of Department of Defense information; Air Force Materiel Command (photo). | GAO-24-106499

**Figure 5.** Fire/Crash Rescue Station Project under Construction.

- Sharing lessons learned. Corps officials told us that after-action reviews and lessons learned are shared at the district level and regional level, but not across the Corps. For example, an after-action review was completed for the Special Operations Forces Training Command Building at Fort Liberty—a project we reviewed during a site visit. However, designers on this project said that they were not using the Corps' database, the Military Mission Lessons Learned System, that could share this project's observations and lessons learned across the Corps. According to reports, the Special Operations Forces Training Command Building at Fort Liberty was delayed for over 3 years due to design errors and according to officials, other factors.[46]

  For example, according to Corps documentation, designers incorrectly assumed a wind load exposure category as outlined in the Structural Engineering UFC, and designers did not include all loading for the building's roof.[47] Design revisions were made in response to design errors, which resulted in increased costs. This included an approximately $2.1 million contract modification for, among other things, revisions for the building's existing structural steel—including beams, columns, and studs. Corps after-action review documentation cites poor coordination during design and a lack of oversight during the quality control process, causing missed

[46] Corps officials said that other factors such as personnel turnover and changing customer requirements also led to schedule delays.

[47] According to documentation, the wind load exposure category should have been designated as a Category C instead of Category B. Also, an analysis of the building's structural system found that roof beams had deflections that exceeded code limits.

opportunities to catch design errors (see fig. 6).[48] The responsible district subsequently took corrective actions to improve its quality control process, including issuing design quality control checklists and revising the quality management plan for projects within the district.

**Special Operations Forces (SOF) Training Command Building**

**Location:** Fort Liberty, NC

**Construction start date:** March 2017

**Original planned construction completion date:** February 2019

**Actual construction completion date:** July 2022

**Original construction amount:** Approximately $38.7 million

**Current obligation amount:** Approximately $54.3 million

**Scope of project:** The SOF Training Command Building is a three-story, 138,355 square foot facility for the John F. Kennedy Special Warfare Center and School. The building includes administrative and operations space, classrooms, and a library.

**Reasons for schedule delays:** The SOF Training Command building's schedule delays are attributed to design errors and insufficient quality control processes, personnel turnover, and changing customer requirements.

Source: GAO analysis of Department of Defense information; U.S. Army Corps of Engineers (photo). | GAO-24-106499

**Figure 6.** Special Operations Forces Training Command Building under Construction.

The Corps has also acknowledged issues with documenting and sharing observations and lessons learned across the enterprise. Corps documentation and officials said that observations and lessons learned are not published in a way that makes them accessible to others. For example, Corps documentation states that the archived database containing after-action reviews and lessons learned for military construction projects was accessed by fewer than 60 users from August 2022 through August 2023.

Army Engineer Regulation 1110-3-12 states that lessons learned from after-action reviews should be documented at the conclusion of each of the four phases of a military construction project and shared regionally.[49] Further, Department of the Army guidance states that lessons-learned program fundamentals include a formal lessons-learned policy letter or standard operating procedure and recurring training for staff, personnel, and students on the lessons-learned program.[50] Additionally, Army Regulation 11-33 states that an effective lessons-learned program prevents mistakes and repeats successes, which leads to saving resources.[51]

---

[48] See U.S. Army Corps of Engineers South Atlantic Division, *Organizational and Process Changes by Savannah District Engineering Division to Address Design Quality Issues.*

[49] Army Engineer Regulation 1110-3-12.

[50] Department of the Army Pamphlet 11-33, *Guide to the Army Lessons Learned Program* (July 28, 2021).

[51] Army Regulation 11-33.

However, the Corps does not have a framework in place that ensures that after-action reviews take place and that lessons learned are documented and shared across the enterprise because it has not issued needed guidance and implemented an operational enterprise-wide lessons-learned system. Additionally, it has not developed training that would be needed to implement an enterprise-wide lessons-learned system.

- The Corps does not provide detailed guidance for how military construction project observations and lessons should be validated, captured, or shared. The Corps concluded in fiscal year 2023 that there is no policy or strategy for capturing lessons learned at the enterprise level, and any observations are captured on an ad-hoc basis. Corps officials stated that they have guidance that requires personnel to document after-action reports and lessons learned. Specifically, Army Engineer Regulation 1110-3-12 discusses the goals of an after-action report and when it is required to be completed— such as when there is a design schedule slippage of 30 days or more.[52] However, the guidance does not provide detail for how to capture and validate project lessons learned from after-action reports or how personnel should share observations and lessons learned at the enterprise level, such as through a standard operating procedure.
- Corps officials stated that they share documents and presentations regarding projects at the district or regional level and within discipline communities of practice. For example, the Louisville district completed an after-action report in October 2022 that includes project lessons learned, successes, and challenges.[53] A Louisville district official said that this after-action report was conducted across the district and was also shared with some regional officials. However, sharing information on an ad-hoc basis with only the district and region does not ensure that all individuals who may need to know will have access. It also does not ensure that project participants have documented guidance for how best to capture and share information.

---

[52] Specifically, the guidance states that after-action reviews must be performed when an innovation has resulted in a significant project success or an error or other significant change causes one or more of these conditions to occur: a cost increase of 5 percent or more, a design schedule slippage of 30 days or more, a construction time growth of 60 days or more, or a consequent reduction in project quality. Army Engineer Regulation 1110-3-12.

[53] U.S Army Corps of Engineers, Louisville District Military & IIS Project Management Branch After Action Review (Oct. 12, 2022.

- The Corps does not have an operational enterprise-wide lessons-learned system that could better enable staff to share and access information across the agency. The current system for housing after-action reviews, the Military Mission Lessons Learned System, is not operational and users are not able to enter lessons learned into the system for vetting. Corps officials told us that they would like to transition this system into a new, more user-friendly system, but that this transition is on hold pending decisions on an enterprise-wide lessons-learned system (discussed below).
- The Corps provides some training to personnel for documenting after-action reports, validating lessons learned, and sharing best practices. However, additional training would be needed if it updates its guidance and implements an enterprise-wide system for lessons learned. Corps headquarters officials also emphasized that personnel need training for reviewing and learning how to vet lessons learned and share best practices.

The Corps is taking steps to improve its program and address these challenges. For example, it is drafting a lessons-learned strategy that is intended to provide a guideline for the Corps to capture and share lessons and best practices in support of future projects' execution.[54] The Corps is also in the process of developing an implementation plan for its strategy. The Corps' draft strategy and implementation plan include information on developing lessons-learned policy, implementing an enterprise-wide system for sharing lessons learned, and training for Corps personnel.

However, officials told us that their plans are in draft, with a potential target of finalizing the draft by fiscal year 2025. Additionally, the draft strategy includes a fiscal year 2027 enterprise-wide implementation milestone. Corps officials told us that a process where Corps employees actively share and access observations and lessons learned in the near term, including a functional database to enable this, would be beneficial and would lead to military project design and construction that is better, faster, cheaper, and safer.

---

[54] Officials told us that as part of their effort to develop a lessons-learned strategy, they reviewed the Corps' lessons-learned process and program with input from senior leaders and officials across the Corps. In addition, the Corps is developing an implementation plan to describe how it will execute its strategy. According to draft documentation, the Corps has a goal of full implementation in fiscal year 2027.

By developing and issuing guidance for documenting after-action reviews and capturing and validating lessons learned, the Corps could help personnel learn how to effectively document these reviews at various stages of a military construction project. Additionally, implementing in an expedient manner an enterprise-wide system to share lessons learned and project best practices, including a functional database, could allow Corps personnel to more consistently use project observations generated from past projects to inform current projects. Finally, developing training could enable the Corps to better implement an enterprise-wide approach to documenting observations and validating lessons learned. As noted above, annual reports to Congress on military construction delays over the previous 5 fiscal years show that approximately 25 percent of delayed projects were due to poor initial planning—which includes factors that could have been reasonably addressed before a delay occurred.

Therefore, a system that could store and share project observations from past projects may help prevent the Corps from committing design errors on current and future projects, which in turn may help prevent schedule delays and cost overruns.

## NAVFAC

NAVFAC officials stated although NAVFAC does not yet have a lessons-learned program, it has taken initial steps to create an enterprise-wide program. For the four projects we reviewed, we found that NAVFAC did not conduct after-action reviews. For example, the design manager of NAVFAC's delayed Road Construction and Repair Project at Camp Lejeune told us there was not an after-action review completed for this project and the project manager was unsure of any formal processes for documenting lessons learned. This project began construction in September 2020 and was originally scheduled to be completed in April 2022. According to NAVFAC officials, it was completed in June 2024.

According to DOD's fiscal year 2022 annual report to Congress on military construction delays and a NAVFAC report, this project was delayed due to design errors and unforeseen conditions, such as underground utilities that were in disrepair (see fig. 7).[55] For example, NAVFAC officials said that

[55] According to a NAVFAC report, a significant amount of rainfall and utility strikes in the area also led to delays.

there was not a survey completed in the initial design with data for utilities. In addition, the design did not include information about a 24-inch water line that was discovered during construction. According to the officials, this resulted in a contract modification of about $170,000 to reinforce the existing water main. NAVFAC officials emphasized that a lesson learned from this project could be to invest more money into quality surveys at the beginning of the project rather than dealing with issues during construction.

**Road construction and repair**

**Location:** Camp Lejeune, NC

**Construction start date:** September 2020

**Original planned construction completion date:** April 2022

**Actual construction completion date:** June 2024

**Original construction amount:** Approximately $17.6 million

**Current obligation amount:** Approximately $18.5 million

**Scope of project:** The project includes road improvements—including construction of a service and access road, widening of an existing road, creation of jogging trails, and installation and movement of utility poles.

**Reasons for schedule delays:** The roadwork's schedule delays are attributed to design errors, unforeseen conditions such as underground utilities that were in disrepair, significant rainfall and utility strikes by contractors.

Source: GAO analysis of Department of Defense information; Military & Federal Construction Company, Inc (photo). | GAO-24-106499

**Figure 7.** Road Construction and Repair Project under Construction.

In addition, construction of NAVFAC's Communications/Crypto Facility at Joint Base Pearl Harbor-Hickam was delayed due to design errors, material delays, and issues with the Direct Digital Control System for the HVAC system. This project began construction in July 2019 and was originally scheduled to be completed in April 2021. As of June 2024, officials said that the project is not yet complete. Also, NAVFAC officials said that there is no after-action review for this project.

According to NAVFAC documentation, design errors that led to schedule delays and cost growth for the project included the failure to provide the Hazardous Material Survey Report and reroofing plans in the final design submission.[56] As a result, contract modifications were made for almost $3.6 million to remove hazardous materials and approximately $2.3 million to repair the roof. However, NAVFAC officials said that it was NAVFAC's responsibility to provide quality assurance over the contracted architecture and engineering firm to ensure that the necessary documents were included in the design. In addition, the project was delayed for 942 days, and a contract modification was made for approximately $4.8 million in July 2019 to replace

---

[56] Naval Facilities Engineering Systems Command, Pacific Memorandum, Contract N62742-13-D-0004, Task Order 009, FY18 Project P-013, Communications/Crypto Facility, Wahiawa, Joint Base Pearl Harbor Hickam, Pearl Harbor, Hawaii; Letter of Concern (Feb. 26, 2020).

the HVAC Direct Control System to meet cybersecurity requirements (see fig. 8).[57] NAVFAC project managers also emphasized that the design for renovation projects can be more complex than new projects and noted that this project also contained a Sensitive Compartmented Information Facility, which contains many technical requirements.

**Communications/Crypto Facility**

**Location:** Joint Base Pearl Harbor Hickam, HI

**Construction start date:** July 2019

**Original planned construction completion date:** April 2021

**Actual construction completion date:** To be determined

**Original construction amount:** Approximately $49.8 million

**Current obligation amount:** Approximately $124.0 million

**Scope of project:** The project consists of renovations to include converting a communications building into a Sensitive Compartmented Information Facility. It also includes replacing existing power and Heating Ventilation and Air Conditioning (HVAC) systems and renovating existing walls, ceilings, and doors to meet secure area requirements.

**Reasons for schedule delays:** The Communication/Crypto facility's schedule delays are attributed to design errors, material delays, and issues regarding an HVAC Direct Digital Controller.

Source: GAO analysis of Department of Defense information; Naval Facilities Engineering Systems Command (photo). | GAO-24-106499

**Figure 8**. Communications/Crypto Facility Project.

Navy and Marine Corps FC state that during various design stages, the quality control review must include, among other things, verifying current criteria and incorporating lessons learned.[58] Additionally, according to Navy guidance on the Navy Lessons Learned Program, Echelon II shore commanders, such as the commander of NAVFAC, should, among other things, designate a lesson manager to oversee the lessons-learned program, standardize and publish lessons-learned processes and procedures, and establish and operate an issue resolution process to rapidly adapt best practices. The Navy guidance also states that Echelon III shore commanders, or the commanders of NAVFAC Atlantic or NAVFAC Pacific, should provide training and guidance to subordinate commands on the collection and reporting of observations following various activities. Finally, it states that

---

[57] NAVFAC issued an instruction in June 2019 that requires the standardization of facility-related controlled systems—including HVAC Control Systems—to improve cybersecurity and increase operational effectiveness, among other things. Commander, NAVFAC Instruction 11000.2, *Facility Related Control System Standardization* (June 3, 2019). In October 2019, NAVFAC Hawaii conducted a business case analysis which recommended the standardization of a HVAC programmable controller from a single manufacturer. NAVFAC officials said that the original design for the project in 2017 did not include plans for this specific type of HVAC programmable controller, which led to delays. However, NAVFAC officials said that the switch to the required HVAC programmable controller will be more cost-effective overall because their public works personnel have the expertise to work on them.

[58] FC 1-300-09N, Navy and Marine Corps Design Procedures (May 17, 2024).

Navy organizations should conduct comprehensive after-action reviews to ensure the Navy learns from its experiences.[59]

A recent Naval Safety Command assessment of NAVFAC's risk processes from June 2023 also found that its processes did not comply with the Navy's Lessons Learned Program.[60] The report states that there is not a lessons-learned program that promotes continual learning across all functional areas. NAVFAC concurred with this finding and has created a corrective action plan. The corrective action plan states that NAVFAC should create lessons-learned processes and procedures by February 2024; and that NAVFAC should institutionalize and train personnel in accordance with these processes and procedures by April 2024.

However, NAVFAC has not met these milestones. Officials told us they are in the initial stages of developing a policy to comply with the Navy's Lessons Learned Program guidance, which may take several months to finalize.

NAVFAC officials stated that they share information regarding projects through technical discipline communities of practice. However, sharing information on an ad-hoc basis does not ensure that all individuals who may need to know will have access. NAVFAC officials also told us that a system where NAVFAC employees actively share and access observations and lessons learned is needed to help close information gaps and to prevent mistakes that could occur again in the future. A NAVFAC official and documentation said that they plan to use the Joint Lessons Learned Information System in the future to share lessons learned, but there are no details on a plan for using the system.[61] A NAVFAC official also said that they need to first assign and conduct training for lessons-learned managers.

NAVFAC does not provide guidance for how project observations and lessons should be captured or shared. In addition, NAVFAC does not provide training or support to personnel for validating lessons learned or sharing best practices. NAVFAC officials agreed that guidance is required for personnel to

---

[59] Office of the Chief of Naval Operations Instruction 3500.37E, *Navy Lessons Learned Program* (Jan. 25, 2024).

[60] Naval Safety Command, *Commander, Naval Facilities Engineering Systems Command Assurance Assessment 23-11* (July 18, 2023).

[61] The Joint Lessons Learned System is the DOD system of record for gathering, developing, and disseminating joint lessons learned for the armed forces. The system facilitates the collection, tracking, management, sharing, dissemination and archiving of information to improve the development, design, and readiness of the joint force. Chairman of the Joint Chiefs of Staff Instruction 3150.25H, *Joint Lessons Learned Program* (Dec. 30, 2021) (incorporating change 1, effective Apr. 5, 2024).

capture project observations lessons learned and that training is needed for personnel to validate lessons learned.

Developing and issuing guidance for after-action reviews and lessons learned could help NAVFAC personnel learn how to effectively document these reviews at various stages of a military construction project. In addition, by implementing a process or mechanism to share observations and lessons learned, NAVFAC could help personnel consistently use observations generated from past projects to inform current projects.

Finally, developing training for personnel could enable NAVFAC to consistently document project observations and validate lessons learned as a further means to inform current projects. DOD's annual reports to Congress on military construction delays over the previous 5 fiscal years show that approximately 25 percent of projects delayed were due to poor initial planning—which includes factors that could have been reasonably addressed before a delay occurred. The use of past project observations may help prevent NAVFAC from committing design errors on current and future projects, which in turn may help prevent schedule delays and cost overruns.

## Conclusion

Proper planning of military construction projects is critical to avoid design errors that could lead to project schedule delays and increased costs.

Within DOD, the significant number of military construction projects delayed at least a year or more underscores the importance of military construction program oversight. The execution of DOD's military construction program relies on general oversight from offices within OSD, and on project oversight by DOD construction agents. However, OSD does not fully monitor the military construction program because it does not collect relevant information to enable effective oversight of all projects, such as the DOD construction agent responsible for the project, planning and design milestones, and details on funds spent. Requiring project sponsors to report more relevant information would better position DOD to conduct oversight and monitor the execution of projects to ensure the most efficient, expeditious, and cost-effective accomplishment of the military construction program.

Moreover, although documenting and sharing lessons learned do not guarantee that future mistakes will not occur, such practices can help organizations prevent mistakes and save resources. However, the Corps and NAVFAC do not consistently document and share military construction project lessons learned. Developing guidance, a system or process for sharing lessons learned, and training could help both the Corps and NAVFAC to use past project observations and lessons learned to inform their current and future military construction projects. Taking these steps may also help prevent the Corps and NAVFAC from committing design errors, which may reduce schedule delays and cost overruns.

## Recommendations for Executive Action

We are making a total of seven recommendations, including one to DOD, three to the Army, and three to the Navy. Specifically:

The Secretary of Defense should ensure that the Assistant Secretary of Defense (Energy, Installations, and Environment) issues guidance requiring project sponsors to report relevant information necessary to monitor the execution of the military construction program. Such information could include the DOD construction agent responsible for the project, planning and design milestones, and details on funds spent. (Recommendation 1)

The Secretary of the Army should ensure that the Commanding General of the U.S. Army Corps of Engineers develops and issues guidance for documenting after-action reviews and validating lessons learned of military construction projects at the enterprise level. (Recommendation 2)

The Secretary of the Army should ensure that the Commanding General of the U.S. Army Corps of Engineers implements in an expedient manner an enterprise-wide system, including a functional database, to share lessons learned and project best practices within the U.S Army Corps of Engineers for military construction projects. (Recommendation 3)

The Secretary of the Army should ensure that the Commanding General of the U.S. Army Corps of Engineers develops training for personnel to capture and validate lessons learned and best practices for military construction projects. (Recommendation 4)

The Secretary of the Navy should ensure that the Commander, NAVFAC develops and issues guidance for documenting after-action reviews and validating lessons learned of military construction projects within NAVFAC. (Recommendation 5)

The Secretary of the Navy should ensure that the Commander, NAVFAC, implements a process or mechanism to share lessons learned and project best practices for military construction projects within NAVFAC. (Recommendation 6)

The Secretary of the Navy should ensure that the Commander, NAVFAC develops training for personnel to capture and validate lessons learned and best practices for military construction projects. (Recommendation 7)

## Agency Comments

We provided a draft of this report to DOD for review and comment. In its written comments, reproduced in appendix IV, DOD concurred with all of our recommendations and stated that it is taking action to implement them.

DOD also provided technical comments, which we incorporated as appropriate.

We are sending copies of this report to the appropriate congressional committees, the Secretary of Defense and the Secretaries of the Army, Navy, and Air Force.

Alissa H. Czyz Director
Defense Capabilities and Management

## Appendix I: Organizations and Military Construction Projects GAO Contacted

To obtain information for our review, we interviewed officials from the following organizations and military construction projects.

- Office of the Secretary of Defense
    - Office of the Under Secretary of Defense for Acquisition and Sustainment, Deputy Assistant Secretary of Defense for Infrastructure Modernization and Resilience
- Department of the Air Force
    - Air Force Civil Engineer Center
- Department of the Navy

- Assistant Secretary of the Navy (Energy, Installations and Environment)
- Naval Facilities and Engineering Systems Command, Engineering Criteria and Programs Office
- Naval Facilities Engineering Systems Command Atlantic
- Naval Facilities Engineering Systems Command Pacific
- Naval Facilities Engineering Systems Command Hawaii
- Communications/Crypto Facility, Joint Base Pearl Harbor, Hawaii
- Waterfront Improvements Wharves, Joint Base Pearl Harbor, Hawaii
- Naval Facilities Engineering Systems Command Mid-Atlantic
- Road Construction and Repair, Camp Lejeune, North Carolina
- Naval Facilities Engineering Systems Command Northwest
- Undersea Vehicle Maintenance Facility, Naval Base Kitsap, Washington
- Marine Corps Installations Command

– Department of the Army

- Assistant Secretary of the Army (Installations, Energy and Environment)
- Headquarters, U.S. Army Corps of Engineers
- U.S Army Corps of Engineers South Atlantic Division
- U.S Army Corps of Engineers Great Lakes and Ohio River Division
- U.S Army Corps of Engineers Baltimore District
- U.S Army Corps of Engineers Savannah District and Wilmington District
- Special Operations Forces Training Command Building, Fort Liberty, North Carolina
- U.S Army Corps of Engineers Louisville District
- Fire/Crash Rescue Station, Wright-Patterson Air Force Base, Ohio
- U.S Army Corps of Engineers Mobile District
- Aircraft and Flight Equipment Building, Redstone Arsenal, Alabama
- U.S Army Corps of Engineers Sacramento District

- Missile Motor Receipt/Storage Facility, Utah Test and Training Range, Utah

– Criteria Working Groups
- Structural Working Group
- Mechanical Working Group
- Architecture Working Group
- Electrical Working Group
- Sustainability Working Group

## Appendix II: Working Groups

According to Department of Defense (DOD) Military Standard 3007G, Department of Defense Standard Practice: Unified Facilities Criteria, Facilities Criteria, and Unified Facilities Guide Specifications, discipline and functional working groups (working groups) are established or eliminated depending on current or projected criteria program management needs.[62] The Coordinating Panel recommends the establishment or elimination of working groups to the Engineering Senior Executive Panel for concurrence and approval. As of March 2024, approved working groups are as listed in Table 2 below.

**Table 2.** Working groups approved by the engineering senior executive panel

| Discipline working group | Functional working group |
|---|---|
| Architecture | Aviation |
| Civil | Comprehensive Planning |
| Cost Engineering | Construction |
| Electrical | Contingency Engineering |
| Discipline working group | Functional working group |
| Fire Protection | Control Systems |
| Geotechnical | Design-Build |
| Mechanical | Facility Space Planning |
| Pavements/Airfields | Fuels |
| Structural | Installation Resiliency |
| | Medical |

[62] Department of Defense Military Standard 3007G, Department of Defense Standard Practice: Unified Facilities Criteria, Facilities Criteria, and Unified Facilities Guide Specifications (Nov. 2019).

| | |
|---|---|
| | Security |
| | Specifications |
| | Sustainability |
| | Waterfront |

Source: Department of Defense information. | GAO-24-106499.

# Appendix III: Incorporation of Relevant National Defense Authorization Act Provisions

**Table 3.** Department of Defense's (DOD) incorporation of ufc-related NDAA provisions, fiscal years 2018–2022

| National Defense Authorization Act (NDAA) provision | Description or excerpt[a] | Incorporation into Unified Facilities Criteria (UFC), other guidance, or processes[b] | Note |
|---|---|---|---|
| National Defense Authorization Act for Fiscal Year 2018, Pub. L. No. 115-91 (2017) | | | |
| (1) Sec. 2872: Modification of Department of Defense Guidance on Use of Airfield Pavement Markings | (a) Modification required.—Except as provided in subsection (b), the Secretary of Defense shall require such modifications of Unified Facilities Guide Specifications (UFGS) for pavement markings (UFGS 32 17 23.00 20, Pavement Markings; UFGS 32 17 24.00 10 Pavement Markings), Air Force Engineering Technical Letter 97–18, Guide Specification for Airfield and Roadway Marking, and any other DOD guidance on airfield pavement markings as may be necessary to prohibit the use of Type I glass beads or any glass beads with a 1.6 refractive index or less from use on airfield markings on airfields under the control of the Secretary.<br>(b) EXCEPTION.—Subsection (a) shall not apply if the Secretary of the Air Force submits a certification to the congressional defense committees that, whenever a proposed contract for airfield pavement markings includes the use of Type I and Type III glass beads, the assessment of the life-cycle costs associated with the use of such beads appropriately considers the local site conditions, life-cycle cost maintenance, environmental impact, operational requirements, and the safety of flight.. . . | ● | Under the statute, the modifications required under subsection (a) were to apply with respect to procurements occurring after September 30, 2018. |
| (2) Sec. 2875: Permitting Machine Room-less Elevators in Department of Defense Facilities | a) In general—The Secretary of Defense shall issue modifications to all relevant construction and facilities specifications to ensure that machine room-less elevators (MRL) are not prohibited in buildings and facilities throughout DOD, including modifications to the UFGS, the Naval Facilities Engineering Command Interim Technical Guidance, and the Army Corps of Engineers Engineering and Construction Bulletin.<br>(b) Conforming to best practices. —In addition to the modifications required under subsection (a), the Secretary may issue further modifications to conform generally with commercial best practices as reflected in the safety code for elevators and escalators as issued by the American Society of Mechanical Engineers. | ● | This provision was implemented in UFGS 14 21 23.<br>DOD officials stated the terms machinery space (MS) and elevator control room (CR) within UFGS 14 21 23 only apply to MRLs. |

**Table 3.** (Continued)

| National Defense Authorization Act (NDAA) provision | Description or excerpt[a] | Incorporation into Unified Facilities Criteria (UFC), other guidance, or processes[b] | Note |
|---|---|---|---|
| (3) Sec. 2878: Report on Hurricane Damage to Department of Defense Assets. | (a) In general.—Not later than 90 days after the date of the enactment of this Act, the Secretary of Defense shall submit to the congressional defense committees a report on damage to DOD assets and installations from hurricanes during 2017.<br>(b) Elements.—The report required under subsection (a) shall include the following elements:<br>. . .<br>(4) An adaptation plan to ensure military installations funded with taxpayer dollars are constructed to better withstand flooding and extreme weather events. | ● | DOD submitted this report in January 2019 and stated that when DOD constructs a new facility or renovates an older facility, structural upgrades, building envelope materials, and service systems are specially selected to help guard against the effects of hurricanes by applying UFC criteria more stringent than International Code published by the International Code Council. DOD cites UFC 3-301-01 Structural Engineering, UFC 4-023-10, Safe Havens, and UFC 4-150-06 Military Harbors and Costal Facilities as examples. |
| John S. McCain National Defense Authorization Act for Fiscal Year 2019, Pub. L. No. 115-232 (2018) | | | |
| (4) Sec. 2805: Updates and Modifications to Department of Defense Form 1391, Unified Facilities Criteria, and Military Installation Master Plans (as amended) | (a) Flood risk disclosure for military construction.—<br>(1) In general.—The Secretary of Defense shall modify DOD Form 1391 to require, with respect to any proposed major or minor military construction project requiring congressional notification or approval—<br>(A) disclosure whether a proposed project will be sited within or partially within a 100-year floodplain or a 500-year floodplain if outside a 100-year floodplain, according to the most recent available Federal Emergency Management Agency flood hazard data, or will be impacted by projected current and future mean sea level fluctuations over the lifetime of the project; and<br>(B) if the proposed project will be sited within or partially within a floodplain described in subparagraph (A) or will be impacted by projected current and future mean sea level fluctuations over the lifetime of the project, the specific risk mitigation plan. | ◐ | DOD officials told us that UFC amendments to address subsection (b) relating to DOD Form 1391 are currently being drafted and project that DOD may be ready to approve such amendments by the end of 2024. |

| National Defense Authorization Act (NDAA) provision | Description or excerpt[a] | Incorporation into Unified Facilities Criteria (UFC), other guidance, or processes[b] | Note |
|---|---|---|---|
| | (2) Delineation of floodplain.—To the extent that Federal Emergency Management Agency flood hazard data are not available for a proposed major or minor military construction site, the Secretary concerned shall establish a process for delineating the 100-year floodplain using risk analysis that is consistent with the standards used to inform federal flood risk assessments. | | |
| | (3) Reporting requirements.—For proposed projects that are to be sited within or partially within a 100-year floodplain or are to be impacted by projected current and future mean sea level fluctuations over the lifetime of the project, the Secretary concerned shall submit to the congressional defense committees a report with the following: | | |
| | (A) An assessment of flood vulnerability for the proposed project using hydrologic, hydraulic, and hydrodynamic data, methods, and analysis that integrate current and projected changes in flooding based on climate science over the anticipated service life of the facility and future forecasted land use changes. | | |
| | (B) Any information concerning alternative construction sites that were considered, and an explanation of why those sites do not satisfy mission requirements. | | |
| | (C) A description of planned flood mitigation measures. | | |
| | (D) A description of how the proposed project has taken into account projected current and future flood risk and mean sea level fluctuations over the lifetime of the project. | | |
| | (4) Minimum flood mitigation requirements.—When mitigating the flood risk of a major or minor military construction project within or partially within the 100-year floodplain or that will be impacted by projected current and future mean sea level fluctuations over the lifetime of the project, the Secretary concerned shall require any mitigation plan to assume— | | |
| | (A) an additional 2 feet above the base flood elevation for non-mission critical buildings, as determined by the Secretary; | | |
| | (B) 3 feet above the base flood elevation for mission critical buildings, as determined by the Secretary; and | | |

**Table 3.** (Continued)

| National Defense Authorization Act (NDAA) provision | Description or excerpt[a] | Incorporation into Unified Facilities Criteria (UFC), other guidance, or processes[b] | Note |
|---|---|---|---|
| | (C) any additional flooding that will result from projected current and future flood risk and mean sea level fluctuations over the lifetime of the project.<br>(b) Disclosure requirements for DOD Form 1391.— Not later than 30 days after the date of the enactment of this Act, the Secretary of Defense shall amend DOD Form 1391 to require, for each requested military construction project—<br>(1) disclosure whether the project was included in the prior year's future-years defense program submitted to Congress pursuant to section 221 of title 10, U.S. Code; and<br>(2) inclusion of an energy study or life cycle analysis.<br>(c) Incorporation of changing environmental condition projections in military construction design and modifications.—<br>(1) Fiscal year 2019. Not later than 30 days after the date of the enactment of this Act, the Secretary of Defense shall amend section 3–5.6.2.3 of UFC 1–200–01 and UFC 1–200– 02 (or any similar successor regulations) to provide that in order to anticipate changing environmental conditions during the design life of existing or planned new facilities and infrastructure, projections from reliable and authorized sources such as the Census Bureau (for population projections), the National Academies of Sciences (for land use change projections and climate projections), the U.S. Geological Survey (for land use change projections), and the U.S. Global Change Research Office and National Climate Assessment (for climate projections) shall be considered and incorporated into military construction designs and modifications.<br>(2) Fiscal year 2020.<br>(A) Amendments required. Not later than 30 days after the date of the enactment of the National Defense Authorization Act for Fiscal Year 2020, the Secretary of Defense shall amend the UFC as follows:<br>(i) To require that installations of the DOD assess the risks from extreme weather and related effects, and develop plans to address such risks.<br>(ii) To require in the development of such Criteria the use of— | | |

| National Defense Authorization Act (NDAA) provision | Description or excerpt[a] | Incorporation into Unified Facilities Criteria (UFC), other guidance, or processes[b] | Note |
|---|---|---|---|
| | (I) land use change projections through the use of land use and land cover modeling by the U.S. Geological Survey; and<br>(II) weather projections—<br>(aa) from the U.S. Global Change Research Program, including in the National Climate Assessment; or<br>(bb) from the National Oceanic and Atmospheric Administration, if such projections are more up-to-date than projections under item (aa).<br>(iii) To require the Secretary of Defense to provide guidance to project designers and master planners on how to use weather projections.<br>(iv) To require the use throughout the Department of the Naval Facilities Engineering Command Climate Change Installation Adaptation and Resilience planning handbook, as amended (or similar publication of the Army Corps of Engineers). | | |
| National Defense Authorization Act for Fiscal Year 2020, Pub. L. No. 116-92 (2019) | | | |
| (5) Sec. 2804: Amendment of Unified Facilities Criteria to Promote Military Installation Resilience, Energy Resilience, Energy and Climate Resiliency, and Cyber Resilience | (a) Amendment required.—<br>(1) In general.—Not later than September 1, 2020, the Secretary of Defense shall amend the UFC relating to military construction planning and design, to ensure that building practices and standards of DOD promote military installation resilience, energy resilience, energy and climate resiliency, and cyber resilience.<br>(2) Considerations and consultations.—In preparing amendments pursuant to paragraph (1), the Secretary of Defense—<br>(A) shall take into account historical data, current conditions, and sea level rise projections; and<br>(B) may consult with the heads of other federal departments and agencies with expertise regarding military installation resilience, energy resilience, energy and climate resiliency, and cyber resilience. | ● | The statute stated that any DOD Form 1391 submitted to Congress after September 1, 2020, shall comply with the UFC, as amended pursuant to this section. |
| (6) Sec. 2805: Modification to Department of Defense Form 1391 Regarding Consideration of Potential Long-term Adverse Environmental Effects | (a) Modification.— | ◐ | DOD officials stated that the military services currently have guidance, and that DOD intends to fund the creation of a new UFC to house this requirement because it does not contain technical design criteria. |

**Table 3.** (Continued)

| National Defense Authorization Act (NDAA) provision | Description or excerpt[a] | Incorporation into Unified Facilities Criteria (UFC), other guidance, or processes[b] | Note |
|---|---|---|---|
| | (1) Certification requirement.—The Secretary of Defense shall modify DOD Form 1391 to require, with respect to any proposed major or minor military construction project requiring congressional notification or approval, the inclusion of a certification by the Secretary of Defense or the secretary of the military department concerned that the proposed military construction project takes into consideration—<br>(A) the potential adverse consequences of long-term changes in environmental conditions, such as increasingly frequent extreme weather events, that could affect the military installation resilience of the installation for which the military construction project is proposed; and<br>(B) building requirements in effect for the locality in which the military construction project is proposed and industry best practices that are developed to withstand extreme weather events and other consequences of changes in environmental conditions.<br>(2) Elements of certification.—As part of the certification required by paragraph (1) for a proposed military construction project, the secretary concerned shall identify the potential changes in environmental conditions, such as increasingly frequent extreme weather events, considered and addressed under subparagraphs (A) and (B) of paragraph (1).<br>. . . | | GAO is tracking progress on DOD's actions, which currently have a target completion date of February 2025. |
| William M. (Mac) Thornberry National Defense Authorization Act for Fiscal Year 2021, Pub. L. No. 116-283 (2021) | | | |
| (7) Sec. 2805: Congressional Project Authorization Required for Military Construction Projects for Energy Resilience, Energy Security, and Energy Conservation | (a) Replacement of notice and wait authority.—Section 2914 of title 10, United States Code, is amended to read as follows:<br>"§ 2914. Military construction projects for energy resilience, energy security, and energy conservation"<br>"(b) (1) Submission of project proposals. As part of DOD Form 1391 submitted to the appropriate committees of Congress for a military construction project covered by subsection (a), the Secretary of Defense shall include the following information:<br>"(A) The project title.<br>"(B) The location of the project.<br>"(C) A brief description of the scope of work. | ● | DOD officials noted they implemented this provision when submitting the relevant forms.<br>The requirements for DOD Form 1391 stipulated in the amended 10 U.S.C. § 2914 applied to certain military construction projects for fiscal year 2023 and thereafter. |

| National Defense Authorization Act (NDAA) provision | Description or excerpt[a] | Incorporation into Unified Facilities Criteria (UFC), other guidance, or processes[b] | Note |
|---|---|---|---|
| | "(D) The original project cost estimate and the current working cost estimate, if different.<br>"(E) Such other information as the Secretary considers appropriate.<br>"(2) In the case of a military construction project for energy conservation, the Secretary also shall include the following information:<br>"(A) The original expected savings-to-investment ratio and simple payback estimates and measurement and verification cost estimate.<br>"(B) The most current expected savings-to-investment ratio and simple payback estimates and measurement and verification plan and costs.<br>"(C) A brief description of the measurement and verification plan and planned funding source.<br>"(3) In the case of a military construction project for energy resilience or energy security, the Secretary also shall include the rationale for how the project would enhance mission assurance, support mission critical functions, and address known vulnerabilities." | | |
| National Defense Authorization Act for Fiscal Year 2022, Pub. L. No. 117-81 (2021) | | | |
| (8) Sec. 2805: Flood Risk Management for Military Construction | . . .<br>(d) Conforming amendment of Unified Facilities Criteria.—<br>(1) Amendment required.—Not later than September 1, 2022, the Secretary of Defense shall amend the UFC relating to military construction planning and design to ensure that building practices and standards of DOD incorporate the minimum flood mitigation requirements of section 2805(a) of the Military Construction Authorization Act for Fiscal Year 2019 (division B of Public Law 115–232; 132 Stat. 2262; 10 U.S.C. 2802 note), as amended by this section.. . . | ● | The statute states that any DOD Form 1391 submitted to Congress after Sept. 1, 2022, is to comply with the UFC as amended by this statute. |

**Table 3.** (Continued)

| National Defense Authorization Act (NDAA) provision | Description or excerpt[a] | Incorporation into Unified Facilities Criteria (UFC), other guidance, or processes[b] | Note |
|---|---|---|---|
| (9) Sec. 2841: Amendment of Unified Facilities Criteria to Require Inclusion of Private Nursing and Lactation Space in Certain Military Construction Projects | (a) Amendment required.—The Secretary of Defense shall amend UFC 1–4.2 (Nursing and Lactation Rooms) of the UFC/DOD Building Code (UFC 1–200–01) to require that military construction planning and design for buildings likely to be regularly frequented by nursing mothers who are members of the uniformed services, civilian employees of DOD, contractor personnel, or visitors include a private nursing and lactation room or other private space suitable for that purpose.<br>(b) Deadline.—The Secretary of Defense shall complete the amendment process required by subsection (a) and implement the amended UFC 1–4.2 not later than one year after the date of the enactment of this Act. | ● | |
| (10) Sec. 2842: Revisions to Unified Facilities Criteria Regarding Use of Variable Refrigerant Flow Systems | (a) Publication and comment period requirements.—The Under Secretary of Defense for Acquisition and Sustainment shall publish any proposed revisions to the UFC regarding the use of variable refrigerant flow systems in the Federal Register and shall specify a comment period of at least 60 days.<br>. . . | ● | In January 2024 DOD updated UFC 3-410-01, Heating, Ventilating, and Air Conditioning Systems, which provides criteria for the use of variable refrigerant flow systems. The UFC states that no revision will be made to it without Federal Register notification. |
| (11) Sec. 2843: Amendment of Unified Facilities Criteria to Promote Energy efficient Military Installations | (a) UFC amendment required.—To the extent practicable, the Secretary of Defense shall amend the UFC relating to military construction planning and design to ensure that building practices and standards of DOD incorporate the latest consensus-based codes and standards for energy efficiency and conservation, including the 2021 International Energy Conservation Code (IECC) and the American Society of Heating, Refrigerating, and Air-Conditioning Engineers (ASHRAE) Standard 90.1-2019.<br>(c) Reporting requirement.—Not later than February 1, 2024, the Secretary of Defense shall submit to the Committees on Armed Services of the House of Representatives and the Senate a report—<br>(1) describing the extent to which the UFC, as amended pursuant to subsection (a), incorporate the latest consensus-based codes and standards for energy efficiency and conservation, including the 2021 IECC and the ASHRAE Standard 90.1- 2019, as required by such subsection; and | ● | The statute stated that the amendment process described in section 2843(a) shall be completed in a timely manner so that any DOD Form 1391 submitted to Congress in connection with a fiscal year 2024 budget submission and thereafter complies with the UFC, as amended by this statute. |

| National Defense Authorization Act (NDAA) provision | Description or excerpt[a] | Incorporation into Unified Facilities Criteria (UFC), other guidance, or processes[b] | Note |
|---|---|---|---|
| | (2) in the case of any instance in which the UFC continues to deviate from such consensus-based codes and standards for energy efficiency and conservation, identifying the deviation and explaining the reasons for the deviation. | | |
| (12) Sec. 2844: Additional Department of Defense Activities to Improve Energy Resiliency of Military Installations | (a) Consideration of including energy microgrid in military construction projects.—<br>(1) Amendment of UFC Required.—The Secretary of Defense shall amend the UFC/DOD Building Code (UFC 1–200–01) to require that planning and design for military construction projects inside the United States include consideration of the feasibility and cost-effectiveness of installing an energy microgrid as part of the project, including intentional islanding capability of at least seven consecutive days, for the purpose of—<br>(A) promoting on-installation energy security and energy resilience; and<br>(B) facilitating implementation and greater use of the authority provided by subsection (h) of section 2911 of title 10, United States Code, as added and amended by section 2825 of the Military Construction Authorization Act for Fiscal Year 2021 (division B of Public Law 116– 283). | ● | |
| (13) Sec. 2881: Clarification of Installation and Maintenance Requirements Regarding Fire Extinguishers in Department of Defense Facilities | Section 2861 of the Military Construction Authorization Act for Fiscal Year 2020 (division B of Public Law 116–92; 10 U.S.C 113 note; 133 Stat. 1899) is amended by striking "requirements of national model fire codes developed by the National Fire Protection Association (NFPA) and the International Code Council" and inserting "NFPA 1, Fire Code of the NFPA and applicable requirements of the international building code and international fire code of the International Code Council." | ● | On 3 February 2020 DOD updated UFC 3-600-01, Fire Protection Engineering for Facilities to state that general purpose portable fire extinguishers must be provided where required by NFPA 101. DOD officials stated that this change had implemented the provision. |
| James M. Inhofe National Defense Authorization Act for Fiscal Year 2023, Pub. L. No. 117-263 (2022) | | | |
| (14) Sec. 2810: Consideration of Installation of Integrated Solar Roofing to Improve Energy Resiliency of Military Installations. | The Secretary of Defense shall amend the UFC/DOD Building Code (UFC 1–200–01) to require that planning and design for military construction projects inside the United States include consideration of the feasibility and cost-effectiveness of installing integrated solar roofing as part of the project, for the purpose of—<br>(1) promoting on-installation energy security and energy resilience; | ● | |

**Table 3.** (Continued)

| National Defense Authorization Act (NDAA) provision | Description or excerpt[a] | Incorporation into Unified Facilities Criteria (UFC), other guidance, or processes[b] | Note |
|---|---|---|---|
| | (2) providing grid support to avoid energy disruptions; and<br>(3) facilitating implementation and greater use of the authority provided by subsection (h) of section 2911 of title 10, U.S.Code | | |
| (15) Sec. 2811: Revision of Unified Facilities Guide Specifications and Unified Facilities Criteria to Include Specifications on Use of Gas Insulated Switchgear and Criteria and Specifications on Microgrids and Microgrid Converters | (a) Gas insulated switchgear.—Not later than one year after the date of the enactment of this Act, the Under Secretary of Defense for Acquisition and Sustainment shall modify the UFGS to include a distinct specification for medium voltage gas insulated switchgear.<br>(b) Microgrids.—Not later than one year after the date of the enactment of this Act, the Under Secretary of Defense for Acquisition and Sustainment shall—<br>(1) modify the UFC to include criteria for microgrids; and<br>(2) modify the UFGS to include specifications for microgrids and microgrid controllers. | ● | The medium voltage switchgear and gas insulation specification was published as UFGS 2613 32 on 1 November 2023.<br>UFC 3-550-04, Resilient Installation Microgrid Design was published 1 March 2024.<br>The microgrids and microgrid controllers guide specifications was formally published in the quarterly UFGS in February 2024. |
| (16) Sec. 2812: Determination and Notification Relating to Executive Orders That Impact Cost and Scope of Work of Military Construction Projects | (a) Determination and update of DOD Form 1391.—Not later than 30 days after the date on which an executive order is signed by the President, the Secretary concerned shall—<br>(1) determine whether implementation of the executive order would cause a cost or scope of work variation for a military construction project under the jurisdiction of the Secretary concerned;<br>(2) assess the potential for life-cycle cost savings associated with implementation of the executive order for such a project; and<br>(3) update DOD Form 1391 for each such project that has not been submitted for congressional consideration, where such implementation would affect such cost or scope of work variation, including—<br>(A) projects to be commenced in the next fiscal year beginning after the date on which the executive order was signed; and<br>(B) projects covered by the future-years defense program submitted under section 221 of title 10, United States Code.<br>(b) Notification to Congress.—Not later than 10 days after determining under subsection (a)(1) that implementation of an executive order would cause a cost or scope of work variation for a military construction project, the Secretary concerned shall submit to the congressional defense committees a report indicating the estimated cost increases, scope of work increases, life-cycle costs, and any other impacts of such implementation. | ○ | DOD officials stated a finalized policy update would be issued by 30 March, 2024.<br>DOD officials explained a separate policy for projects in the planning phase does not have a timeline for publication. |

Legend:

● = The provision has been incorporated.

◐ = The provision has been partially incorporated.

○ = The provision has not been incorporated.

Source: GAO analysis of NDAA and DOD documentation. | GAO-24-106499

[a]Portions of statutes not relevant to UFC modifications have been omitted from this table.

[b]Our review did not assess whether DOD implemented these statutes within time frames that may have been required by each statute.

# Appendix IV: Comments from the Department of Defense

ASSISTANT SECRETARY OF DEFENSE
3400 DEFENSE PENTAGON
WASHINGTON, DC 20301-3400

ENERGY, INSTALLATIONS, AND ENVIRONMENT

August 23, 2024

Ms. Alissa Czyz
Director, Defense Capabilities and Management
U.A. Government Accountability Office
441 G Street NW
Washington, DC 20548

Dear Ms. Czyz:

This is the Department of Defense response to the GAO Draft Report, GAO-24-106499, "MILITARY CONSTRUCTION: Better Information Sharing Would Improve DoD's Oversight," dated July 19, 2024 (GAO Code 106499). The Department of Defense proposed response to the subject report is enclosed. My point of contact is Mr. Michael Rosenblatt, at michael.a.rosenblatt8.civ@mail.mil or (703) 571-9079.

Sincerely,

OWENS.BRENDAN.M.1030451844 Digitally signed by OWENS.BRENDAN.M.1030451844 Date: 2024.08.23 16:12:46 -04'00'

Brendan M. Owens

Enclosure:
As stated

cc:
WHS GAO Affairs Division

**GAO DRAFT REPORT DATED JULY 19, 2024**
**GAO-24-106499 (GAO CODE 106499)**

**"MILITARY CONSTRUCTION: BETTER INFORMATION SHARING WOULD IMPROVE DOD'S OVERSIGHT"**

**DEPARTMENT OF DEFENSE COMMENTS**

**RECOMMENDATION 1:** The Secretary of Defense should ensure that the Assistant Secretary of Defense (Energy, Installations, and Environment) issues guidance requiring project sponsors to report relevant information necessary to monitor the execution of the military construction program. Such information could include the DOD construction agent responsible for the project, planning and design milestones, and details on funds spent.

**DoD RESPONSE:** Concur. The Department sees value in common military construction (MilCon) terminology and updating reporting requirements to facilitate oversight of component military construction programs. DoD is currently developing policy to standardize MilCon terminology, an initial step required to provide a foundation for future enhanced MilCon information sharing.

It's also important to note that DoD conducts oversight at the project level on high risk and high interest projects and programs through a host of portfolio specific briefings and meetings that require components to gather substantial project level data including for the Indo Pacific, Europe, and the Energy Resilience and Conservation Investment Program and high profile individual projects such as the Air Force's Joint Intelligence Analysis Center and Presidential Aircraft Replacement Hangar to name a few. DoD captures relevant information to conduct oversight through these briefings.

**RECOMMENDATION 2:** The Secretary of the Army should ensure that the Commanding General of the U.S. Army Corps of Engineers develops and issues guidance for documenting after-action reviews and validating lessons learned of military construction projects at the enterprise-level.

**DoD RESPONSE:** DoD concurs with this recommendation and agrees with the importance of institutionalizing lessons learned processes within the DoD Construction Agents.

**RECOMMENDATION 3:** The Secretary of the Army should ensure that the Commanding General of the U.S. Army Corps of Engineers implements in an expedient manner an enterprise-wide system, including a functional database, to share lessons learned and project best practices within the U.S. Army Corps of Engineers for military construction projects.

**DoD RESPONSE:** DoD concurs with this recommendation and the benefit of a consistent mechanism for sharing lessons learned information. It's important to note that USACE currently documents and shares lessons learned through several means including but not limited to:

- Community of Practice meetings. This includes regular short form meetings around an hour briefly covering lessons learned, which typically occur monthly or quarterly, and long form meetings covering lessons learned in detail, which occur over multiple days occurring annually or biannually.
- Tri-service Discipline Working Group meetings through the UFC Program
- Facility standardization team meetings, which leverage the USACE Centers of Standardization to provide feedback to the Army Standards.

**RECOMMENDATION 4:** The Secretary of the Army should ensure that the Commanding General of the U.S. Army Corps of Engineers develops training for personnel to capture and validate lessons learned and best practices for military construction projects.

**DoD RESPONSE:** DoD concurs with this recommendation. DoD concurs with this recommendation and the benefit of reviewing and revising existing training for documenting after action reviews and the validating and sharing of lessons learned in military construction projects.

**RECOMMENDATION 5:** The Secretary of the Navy should ensure that the Commander, Naval Facilities Engineering Systems Command (NAVFAC) develops and issues guidance for documenting after-action reviews and validating lessons learned of military construction projects within NAVFAC.

**DoD RESPONSE:** DoD concurs with this recommendation and agrees with the importance of institutionalizing lessons learned processes within the DoD Construction Agents.

**RECOMMENDATION 6:** The Secretary of the Navy should ensure that the Commander, NAVFAC implements a process or mechanism to share lessons learned and project best practices for military construction projects within NAVFAC.

**DoD RESPONSE:** DoD concurs with this recommendation and the benefit of processes that share lessons learned on military construction projects.

**RECOMMENDATION 7:** The Secretary of the Navy should ensure that the Commander, NAVFAC develops training for personnel to capture and validate lessons learned and best practices for military construction projects.

**DoD RESPONSE:** DoD concurs with this recommendation and the benefit of reviewing and revising existing training for documenting after action reviews and the validating and sharing of lessons learned in military construction projects.

# Index